by Bill Nye

Drawings by Terry Marks

Photographs by Tom Owen

A TVbooks, inc. Production

Addison-Wesley Publishing Company

Reading, Massachusetts Menlo Park, California New York
Don Mills, Ontario Wokingham, England Amsterdam Bonn
Sydney Singapore Tokyo Madrid San Juan
Paris Seoul Milan Mexico City Taipei

Many of the designations used by manufacturers and sellers to distinguish their products are claimed as trademarks. Where those designations appear in this book and Addison-Wesley was aware of a trademark claim, the designations have been printed in initial capital letters (e.g., Joy). Science Guy is a registered trademark of Bill Nye.

Neither the Publisher nor the Author shall be liable for any damage which may be caused or sustained as a result of conducting any of the activities in this book without specifically following instructions, conducting the activities without proper supervision, or ignoring the cautions contained in the book.

Library of Congress Cataloging-in-Publication Data

Nye, Bill.
Bill Nye the science guy's big blast of science / by Bill Nye the science guy; drawings by Terry Marks; photographs by Tom Owen.
p. cm.
Includes index.
ISBN 0-201-60864-2
1. Science—Experiments—Juvenile literature. 2. Science—Methodology—Juvenile literature. [1. Science—Experiments. 2. Science—Methodology.] I. Marks, Terry, ill. II. Title.
Q164.N9 1993
507.8—dc20 93-11221
CIP
AC

Designed by Sharman Culley
Edited by John Bell
Production Coordinated by Tom Crouch
Produced by TVbooks, inc.
8513 SE 68th Street
Mercer Island, WA 98040

1 2 3 4 5 6 7 8 9-MA-9796959493
First printing, October 1993

Table of Contents

Acknowledgments

I've run across a lot of people in my life that gave me the enthusiasm for science and learning that made it possible for me to write this book. I'd like to thank my parents (no kidding) and sister for emphasizing education. It really is fun. And my older brother is the guy who made science fun from before I can remember.

I had some terrific teachers, too: Mr. Lawrence, Mr. Flowers, Mr. Cross, Mr. Lang (hey, I can call him "George" now), and Professor Sagan. Also, I'd like to thank my friend and colleague Phil Haldeman for his great insights into critical thinking and quality of evidence.

Thanks, everyone!

THE ROAD TO KNOWLEDGE

chapter one

Suppose you were in a shipwreck, and you washed up on a deserted beach somewhere. Some place you don't know anything about. You don't even know where you are. What would you do? Get food? Get help? Get out of there? You'd think of something...and that thinking would probably be science. No kidding!

What is science anyway? Why study it? What's the big deal? Well, "**science**" comes from the old Latin word that means "**knowledge**." So science is gaining knowledge. Learning about the world and keeping track of what we've learned is what we humans do all the time, whenever we think. This includes you. Our knowledge makes us able to live almost anywhere on Earth, including deserted beaches.

You have to admit that compared to other animals, we humans don't have that much going for us. We don't run as fast as horses. We aren't as big as bears. When it comes down to it, most of us can't even keep warm enough to walk around without all kinds of clothes on. So how come humans are living almost everywhere on our planet? How come we're the only living beings that build factories to make things for ourselves, planes to move ourselves around, and farms to grow our food? How come we're the only form of life that we've ever seen off the planet Earth? Well, it's because deep down, we're scientists—all of us.

Every day each of us tries to figure out what's going to happen next, and how soon. When you're riding a bike, paddling a boogie board, or piloting a rocket, you want to know about roads, waves, or space so that you can ride correctly. If you're making your breakfast, you want to learn how to toast the bread without burning it. If you're playing basketball, baseball, or soccer, you want to learn the best way to judge where a ball is going to land, so that you can jam it, catch it, or kick it. **Everyone's a scientist!**

The Scientific Method

In a way, science is how we handle every question in our lives. It's not just how to do things. Science is the way we **figure out** how to do these things. Scientists, maybe scientists like you, call this way of doing things "the Scientific Method." Science means "knowledge," and "method" means "road" in Greek. So the Scientific Method is the "Road to Knowledge."

By the way, do you know who invented the Scientific Method—who invented science? Well, we did. Humans! It wasn't any one person. It's a process, a path, a road to learning that we humans have come up with. Not bad.

Suppose you were an ancient cave guy or gal, and you wanted to know how to make a fire burn. You might have tried burning rocks, dirt, and then wood. Well, you probably would have figured out that wood works best. Then you might have done some thinking about dry wood instead of wet wood. Dry wood is your best bet.

Looking for things to eat? You might have watched what other animals ate, and then tried it for yourself. See, if you were the shipwrecked sailor or an ancient cave person, you had to watch the world, come up with things to try, and then try them. That's using the Scientific Method.

See, every human on Earth should be a scientist. Every human should use **reason** and **knowledge**. Whether you're shopping for a new bike, going after the next level in a video game, or trying to find a cure for a deadly disease, scientific thinking will help you.

Apparently, it's part of human nature to try to figure things out. This quest for knowledge allows us to predict the future and be ready for it. Strange as it may seem, this kind of thinking affects everything we do. If you're designing airplanes, naturally you'd use science. But science reaches farther than that. If you carry an umbrella one day, even though it's not raining when you leave, you're anticipating rain by using your knowledge of the weather. You're using science. It's cool.

Observation

The Scientific Method starts with looking at the world around us. This is called "observation." It doesn't mean just looking with our eyes. We observe with our ears, our nose, our hands, all parts of our bodies. We also listen to what other people have to say about what they've observed. And scientists have invented devices that let us observe things we can't see, or can't see directly—devices like microscopes, thermometers, clocks, rulers, X-ray machines, telescopes, satellite cameras, electric meters, scales, and measuring cups. Those are the tools of science.

Imagine that you see red and green lights moving slowly in the night sky. You're making an observation. You see where the lights are, you see what color they are and how fast they move. The obvious question is, What could those lights be?

Hypothesis

After we see something, we usually want to know what causes it. So we think of a cause. This is a crucial idea in science. It's called making a "hypothesis" [hie-PAHTH-ih-siss]. A hypothesis involves developing an idea or set of ideas that might explain our observations. When we come up with a hypothesis, we're coming up with what we think is the reason something happened.

WOW!

The word "hypothesis" comes from the Greek words for "the idea below," the underlying idea. "Hypo" means "below," and "thesis" means "idea." Be careful not to confuse "hypo" with "hyper," which means "above." (You may have heard of a "hypodermic" [HIE-poe-derr-mick] needle; it goes below the skin.) So we start with the hypothesis, the idea underneath it all. Maybe it's right; maybe it's wrong; but it looks good so far.

& Induction Deduction

The easiest way to come up with a hypothesis is to use "inductive reasoning." That means leading yourself to the answer from what you already know. A British nobleman named Francis Bacon proposed inductive reasoning as the way to go in 1620. People had been thinking this way before then, but Bacon wrote about tying induction to the Scientific Method.

You've probably heard of the famous "deductions" of Sherlock Holmes. He figured out mysteries just by looking at them—by using deduction. He was very successful. But remember, ol' Sherlock was not a real person. Of course everything worked out for him; he was the star of the Sherlock Holmes stories. If he didn't win, the author would have had to invent someone else for every story. That would have been lame.

In Latin, "deduct" means "to lead from." Deductive reasoning is to take away everything that can't be true, so whatever is left must be the truth. Just like you deduct numbers; you "take away." However, deductive reasoning doesn't always work in the real world because there are so many possibilities that it's very difficult to eliminate everything. For instance, imagine that you found a note in your locker at school. Did your friend put it there? If you use pure deductive reasoning, you have to eliminate everyone else in the entire world—more than five billion people. Phew! It makes much more sense to lead yourself to a likely answer by what you already know. Who'd want to give me a note? Whose handwriting does this look like? You could probably figure out that the note-writer was your friend. Coming up with that hypothesis seems pretty easy, doesn't it?

Let's go back to those red and green lights in the sky. You might think, "It's probably an airplane." You don't go, "Oh, it must be a flying pickle." After all, you know from having observed pickles before that they don't fly (unless someone throws them), and airplanes do. Airplanes also usually have lights on them; pickles don't, usually. That's using inductive reasoning to come up with a hypothesis about those flying lights.

Philosophy

The Scientific Method is kind of new as far as human history goes. The ancient Greeks believed that knowledge of anything could be gained by hypothesizing alone. They believed, in a sense, that if you just sat around thinking about stuff long enough, you could completely understand it. Aristotle was among the most famous of these Greek "philosophers" [fill-OSS-uh-ferz]. In Greek that means "lovers of wisdom."

Aristotle made a great many brilliant observations of his world; he even realized that dolphins were air-breathing animals and not really that much like fish at all. (No one else realized this for about 2000 years.) But Aristotle also was a theoretical guy, who preferred pure reasoning to experimenting. For example, Aristotle had a hypothesis that heavy things fall faster than light things. He used a rock and feather to prove his point. He never tried different sized rocks. Apparently, Aristotle didn't think he needed to. Anyway, as you may know this conclusion was wrong—way off. But people accepted it for about 2000 years.

The ancient Greek philosophers were brilliant guys. They invented logic, for cryin' out loud. Today, we use logic to run modern computers! And the ancient Greeks seemed to be so right and reasonable about so many things that it took other humans centuries to let go of their ideas. See, the problem is that just thinking doesn't always lead to the right answers. We've got to test our thinking. Aristotle just wasn't into experimenting the way we modern scientists are.

LOGIC? Yep, that was my idea...

Experiment

The next step in the Scientific Method is to come up with a test to prove, or disprove, our hypothesis. This test is called an "experiment." We try our experiment and see if what actually happens is what we predicted would happen. If it is, we try to predict what happens in similar situations.

What makes a good experiment? It's a test which gives **clear results**, one way or the other. Either dry wood burns well, or it doesn't. Either a particular medicine increases a person's chances of getting over a disease, or it doesn't. A good experiment is **controlled**: outside things can't interfere with what you're trying to test. And, finally, a good experiment can be done over again by someone else, to check the results. It's **repeatable**.

How could you test those lights in the sky? Well, it's difficult to get close to them, but you can observe them from the ground for a few minutes longer. If the lights come from an airplane, they'll probably continue moving in a straight line at a fast speed, so you can watch if they do that. And you might listen for the sound of an engine. Those observations mean that your hypothesis seems right.

Refinement

But what if those lights in the sky suddenly changed direction? What if they move slowly and you can't hear a thing? Then you have to come up with another hypothesis. Back to the flying pickle? No, think of something else that fits your experience and your new observations. Maybe those lights come from a helicopter, or a blimp. You have to refine your hypothesis; that means change it a little to explain your new observations.

Usually after an experiment, you have to go back and change your hypothesis slightly to get things closer to the results of your experiments. You can do this over and over and get better, more accurate predictions. It makes sense. We all do it all the time. That's how you learn how much ice you like in your drink; how far you can run without getting tired; and how hot you like to have your bath. You try your first idea and refine it.

Knowledge

EVERYBODY

The great thing about the Scientific Method is that a scientist can make a prediction and can come up with a theory, and then other scientists can check it. They don't have to take anybody's word for anything. The knowledge, if it is really true, exists outside of us. We don't have to be there to see it. Someone else can see it, can make the same **observations**, can try the same **experiments**, and see if he or she comes up with the same **results**. If he or she doesn't, well, then another scientist can try these same things and see what happens. The knowledge—the science—exists whether or not we're there.

The fact that science works for anyone is one of the great things about doing things scientifically. You can say, "You don't believe me? Okay, *you* try it." Don't take my word for it. Don't take Aristotle's. You try it and see what you get. Scientific knowledge and phenomena are repeatable. Scientists can repeat the experiment of another scientist, and if they get the same result, well, that's a great step toward being sure of something.

Many ancient people believed that the Earth was flat—even many scientists. But there were experiments that showed some scientists it was round; we'll get to those in Chapter 9. Nowadays, we can check the shape of the Earth in a second. Just get a picture of the Earth from space. Sure enough, the Earth is round and it's about 41,000 kilometers (25,000 miles) around. Pretty cool.

See, science is **self-correcting**. If one theory is wrong, we can change it. We can refine it. Or, if need be, we can just toss it out and start over. We can move ahead and try to figure things out. It's our nature. And so far, the Scientific Method is the best way we've come up with. Go wild, fellow scientists.

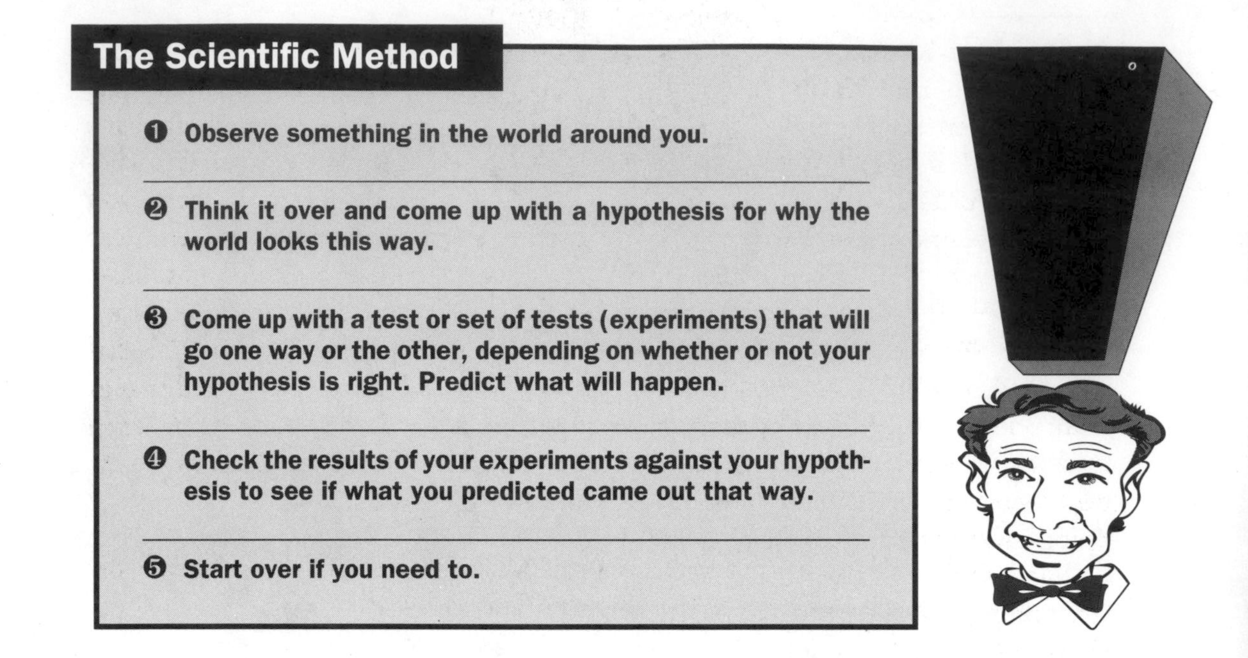

The Scientific Method

1. Observe something in the world around you.
2. Think it over and come up with a hypothesis for why the world looks this way.
3. Come up with a test or set of tests (experiments) that will go one way or the other, depending on whether or not your hypothesis is right. Predict what will happen.
4. Check the results of your experiments against your hypothesis to see if what you predicted came out that way.
5. Start over if you need to.

Here are some more rules we scientists have developed that make life easier for us. These are very useful for dealing with way-out claims like UFOs, the Loch Ness Monster, and being able to predict the future.

Occam's Razor

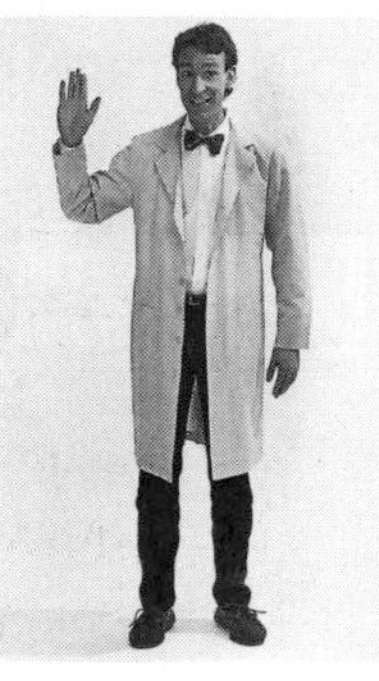

Suppose that I told you that I weigh 5000 pounds. "What?!" you might say—or scream. Because you can see from my pictures that I look like a regular guy. People who look like me don't weigh even close to 5000 pounds. Besides, if I weighed 5000 pounds, I'd probably fall right through the floors of most houses. I'd crush any car I sat in. But then I offered to bet you $50 that I can prove I weigh 5000 pounds. You wouldn't pass up a chance for easy cash, would you?

So we've got a bet. Suppose I bring a scale, like a bathroom scale, and I stand on it. And it reads 5000 pounds. Would that prove that I weigh that much? No. No, no, no. I brought the scale; I could have tampered with it. It's not proof enough. Not even close. Me actually weighing 5000 pounds is only one explanation for the scale reading that high. And that explanation requires so many other way out problems to be solved that it's just not reasonable at all. When people make weird claims, feel free to *not* take them at their word unless they show you proof that doesn't have any simpler explanation.

This line of thinking is connected with an idea that scientists like you and I call "Occam's Razor." William of Occam was a British philosopher who pointed out that if you have a bunch of explanations for some phenomenon, the **simplest explanation** is probably the right one. The idea of a razor is that scientists should "shave" their ideas down to the simplest form. Now we can't prove Occam's Razor as a law or anything. It just makes sense. It also makes life a lot easier.

So here I am standing on a scale that I brought with me to settle our bet. And it says I weigh 5000 pounds. Are you going to say, "Wow, I can't believe it, but your body must be made of something that's never been seen on Earth before, and that's how you weigh so much. Boy, it sure was an unlucky break for me to have made a bet with the only super-heavy science guy that ever lived. Oh well, I lost $50"? No way! Think of different ways to explain what you see, and then use Occam's Razor and choose the simplest explanation.

How many ways can you explain the scale reading?

- I *do* weigh 5000 pounds.
- I weigh 150 pounds, but I have 4850 pounds of rocks in my pockets.
- The gravity right under the scale is stronger than gravity anywhere else on Earth.
- I fooled with the scale so I could get your fifty bucks.

Now which explanation, do you think, is the simplest? Remember, the right answer could be worth $50 to you.

Universality

So you tell me that you think I fooled with the scale and I don't really weigh 5000 pounds. You want to weigh me on your scale, which is blue. "Sorry," I say, "but blue scales don't work on me. See, that color radiance sets up an inverse field that affects my mass anti-gravitationally." (That's just gibberish; ignore it.)

If I really weigh 5000 pounds, I should weigh that much on my scale, your scale, or your next-door neighbor's scale. Or if there is a good reason that I weigh different amounts on different scales, that reason should be testable. I should be able to show you by a new experiment that I'm right. For instance, you can say, "Okay, if blue scales don't work, let's weigh you on my neighbor's white scale instead." Or you can say, "Why don't I step on your scale and see if it tells me the same number as my scale this morning?"

See, one amazing thing about science is that it works the same everywhere. If I weigh 5000 pounds on my scale, then I should weigh that much on every scale. If the scale gives the right weight for me, it should give the right weight for you, too.

Extraordinary Claims, Extraordinary Proof

And that brings up a useful scientific rule about extraordinary claims, like me weighing 5000 pounds. If I make a claim that goes opposite to your ordinary experience, it's up to me to prove that my claim is true. And the only way for me to do that is to design an experiment which can't be explained any simpler way.

Take another extraordinary claim. Say I come to you and tell you that 150 million years ago, there were animals walking around on this planet that were as big as ten elephants. And that the only reason we don't see these creatures today is because they all died out mysteriously about 65 million years ago. That's an extraordinary claim, isn't it? But I've got extraordinary proof, and you can see it in almost any science museum: fossilized dinosaur skeletons! See, there's no simpler scientific explanation for those fossils than for there to have been dinosaurs like *Brachiosaurus* living millions of years ago.

Or say I told you that even though it looks like the Sun goes around the Earth, actually the Earth goes around the Sun. Now you can just track the Sun over the course of one day, and it looks like the Sun moves through the sky. For thousands of years almost everybody believed that was the way the universe worked. So when a guy named Copernicus suggested that the Earth went around the Sun, everybody demanded extraordinary proof. Scientists managed to come up with that proof: you can read all about it in Chapter 9.

Imagine I tell you that everything you see is made of different combinations of 92 little particles. Extraordinary? Well, you can read about some of the proof in the next chapter. Or what if I tell you that a TV picture is actually blinking on and off 60 times every second? Try Chapter 7.

People are always coming up with extraordinary claims, such as that the continents we're on are moving, or that a particular shampoo can help you get dates, or that a certain kind of mold can cure infections, or that Bigfoot is stomping around the woods near my house. What makes you and me scientists is that we demand extraordinary proof for those claims. We want people to measure the continents movements (they did measure; the continents move a few centimeters each year). We want proof that mold can cure infections (it does; mold makes penicillin). So we should also ask for proof that shampoo can help people get dates or that Bigfoot exists, and nobody's come up with strong proof for either of those.

Humans love myths, you know. We have Halloween and goblins, we have leprechauns and Bigfoots, and we have UFOs. It could be we just have a need for having these myths in our memories. It keeps our minds working, perhaps. But that's no reason to accept some way-out claim without some pretty good way-out proof.

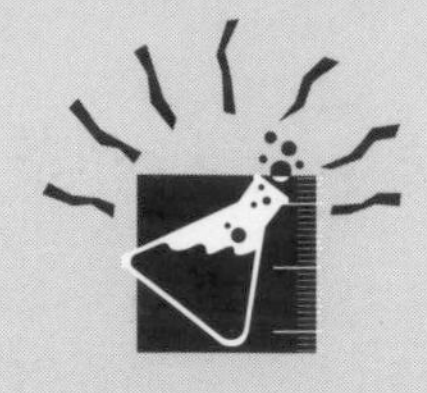

Astrology

I'll give you the classic example of a way-out claim, and then you can test it using the Scientific Method. Get the astrological forecast, or horoscope, out of a newspaper. Many newspapers print one every day. Read all the forecasts. Does the one for your birth sign fit you? Make an observation. Hypothesize about the horoscope. For instance, "This horoscope could fit anybody, whatever his or her birth sign is." Or, "This horoscope only works for Aquariuses." Then design an experiment to test your hypothesis.

How can you test the horoscope? Here's one way. Gather a group of friends who have different astrological signs: Capricorn, Pisces, Gemini, and so on. Take a newspaper or magazine with a horoscope that nobody has read before. Read the predictions out loud, but don't tell your audience which prediction is for which sign. Let them try to match up each prediction with the right person, writing their answers down on a piece of paper. Tally the answers up.

Your friends can almost never match them all up right. They have a one in 12 chance on each one, so they might get lucky, but keep trying this experiment over a few days. You'll see pretty quickly that astrology just doesn't work.

Here's an interesting thing about the astrological signs. They're based on where the stars appeared centuries ago, when astrologers first wrote about them. But the stars have shifted a little. (Read about how things in the universe move in Chapter 9.) Yet almost all astrologers still use the same positions of the stars as centuries ago. None of it really makes any sense. The stars are billions upon billions of kilometers from here. They don't decide our lives for us, but they do fill us with scientific wonder. That part is great.

SCOPING THE STARS

Astrology says that we can figure out what's going to happen and how people behave from where the stars seem to be in the sky. Every year there are more astrology books printed than astronomy books. That's because people want to believe in astrology. People want to know what's going to happen to them so much that they don't use the Scientific Method, Occam's Razor, or the rules of extraordinary proof.

The Science Solution

We are living in a very, very exciting time. Humans may be on the verge of discovering some astonishing things about the workings of the universe. The Scientific Method has been figured out, and we are able to answer so many questions that were not even imagined by our ancestors. But as you may know, we are facing tremendous problems here on the Earth.

The Earth is like that deserted beach we started on at the beginning of this chapter. We have to find food and shelter, find out where we are, find out how to go new places. We have to solve our problems. We have the tools: our brains, our knowledge, and the scientific approach to understanding the nature of the world. So here we go down the road to knowledge.

MATTER

chapter two

Let's start out by talking about everything. That's right, every *thing*. Scientists (like you and me) call every *thing* "matter." Matter is anything that takes up space. We're talking about air, boxes, cars, dust, elephants, figs, gasoline, helium, ink, jeeps, kites, limousines, mosquitoes, notebooks, oxygen, planets, quarks, rockets, sandals, thimbles, underwear, volcanoes, wax, xylophones, you, and zebras. They're usually things that you can see, or feel, or lift. They're all made of some material or groups of materials. In fact, you can see that the words *mater*ial and *matter* come from the same root, an old Latin word meaning "stuff."

What isn't matter? Ideas and feelings (wishes, headache) aren't matter. Directions and measurements (up, one mile) aren't matter. Actions (swimming, wars) aren't matter. Energy (light, TV signals) isn't matter. Changes aren't matter, though when we see a change in the world around us, we're usually seeing a change in matter: iron rusts, water evaporates into the air, the atmosphere turns blue and then cloudy.

Matter	Not Matter
Boats	Swimming
Spilled milk	Thirst
Stars	Wishes
Lasers	Light
Popcorn	Eating
Escalators	Up or Down
Bombs	Wars
Aspirin	Headaches
Snow	North
Television sets	Television signals

Why does matter matter? Well, you and I are made of matter, for one thing! So's our food and water and the air we need to stay alive. So's this book. So's the chair you're sitting on, or the ground you're standing on. Matter is everything everywhere all the time. No wonder we want to know all about it.

Just by looking around, you can probably see three different **phases** [FAY-zez] of matter. Start with this book, or your shirt, or the floor underneath you. They're all **solids**.

Then think about the juice you had with breakfast, or the puddles outside, or the spit in your mouth. These are **liquids**.

Look at the air between you and this book. Actually, you can't see air because it's transparent, but you can feel it when you blow on your hands. Air is a **gas**. The inside of the Sun is a type of gas called "plasma" [PLAZ-muh]. Plasma is gas with so much energy that the outside particles of its atoms are jolted off and electricity passes through it very easily. More on those particles later!

Why does a particular bit of matter come in one phase or another? It depends on how that particular type of matter behaves and how much energy it contains. Here's how to see three phases of the fascinating matter we call "water." Experiments with water are great because our planet Earth is 71% covered with water.

experiment

Solid to Liquid to Gas

Here's What You Need:

- *Zipper-closed plastic bag*
- *4 ice cubes*
- *Microwave oven*

Here's What You Do:

1. *Place four ice cubes inside the plastic bag. Put the bag into the microwave oven.*
2. *Turn the oven on high. Watch through the window.*
3. *When you see the plastic bag start to swell, turn the oven off. Otherwise, the bag could make a major mess inside.*

Here's What You See:

The solid ice melts into liquid water. Then the liquid becomes a gas—water vapor. You can tell that the liquid is changing into a gas because the bag will start to blow up like a balloon. What made the matter change from solid to liquid to gas? The microwave oven gave it more energy. You can see this added energy in how the matter changes, and you can feel it as heat.

Notice how much more room a gas takes up than a solid or a liquid. Gas molecules take up a lot more space because they have a lot more energy. The gas in the bag is water vapor, sometimes also called steam. (The word "steam" is tricky because we also use it to describe *liquid* water droplets so tiny they float in the air, such as the cloud coming out of a boiling kettle or our breath on a cold day.)

Be careful with hot water vapor! It can scald you very fast.

Liquids and Solids

Hold a water balloon in one hand, and a sponge ball in the other. They both have fairly definite shapes, right? Squishy, but pretty much round.

Now toss the water balloon and the sponge ball up in the air. (Did I mention that you should try this experiment outside on a warm day? Too late now!) What happens when the water balloon hits the ground? The rubber breaks. The water goes everywhere. It's not in a round shape anymore—it's spread out flat on the ground. What happens to the sponge ball? It's still round.

This shows the difference between a liquid and a solid. The liquid water takes the shape of its container, the balloon. When that container falls apart, the liquid goes everywhere! The ball made of sponge rubber, a solid, has a shape of its own. It doesn't need a container to keep that shape.

Both liquids and gases are what we call "**fluids**," which means they take the shape of whatever contains them. Hey! What about the air we breathe? There's no container to hold that air, so what keeps it close to the Earth? Actually the air is contained, in a way. Gravity, a powerful force that affects all matter, holds the air in a ball around the Earth. Without gravity, the air would just drift off into space. Seems wild? Well, it is.

All Kinds of Matter

So matter shows up in different phases. It also shows up in more than a million different forms called "**chemicals**." Chemicals aren't just the artificial colors in your breakfast cereal; *all* matter is one chemical or another. There are millions of different chemicals in the universe. We're made of thousands of different chemicals ourselves. And each chemical, each different form of matter, behaves differently. Those differences are important—make that *very, very* important.

Have you ever put salt in the sugar bowl as an April Fool's joke? Ha, ha, ha. You can make your family pretty mad. Salt on cereal—Blechh! You are just sooo funny. Salt may look like sugar, but salt and sugar affect your tongue differently because they're different chemicals. How else can you tell them apart?

Chemical Differences

Here's What You Need:

- *Salt*
- *Sugar*
- *Strong magnifying glass*
- *Candle*
- *Matches*
- *Aluminum foil*
- *A clothespin or tongs*

Here's What You Do:

1. *Use the magnifying glass to look at salt and sugar grains up close. They're pretty cool.*
2. *Form two small dishes out of aluminum foil. Put 4 mL (1/4 teaspoon) of salt in one foil dish, and 4 mL (1/4 teaspoon) of sugar in the other.*
3. *Using the clothespin or tongs, hold the dish of sugar over the candle flame. Wait until it changes.*
4. *Now set the sugar dish down where it won't damage the table, and hold the dish of salt over the candle flame. Does it change?*

Here's What You See:

When you look at the crystals through the magnifying glass, you see that salt and sugar are about the same color. But sugar crystals are bigger than grains of salt. Also, sugar crystals are more jagged; they aren't square like salt crystals. Right away that might make you think that they're different chemicals.

When you heat the crystals over the flame, the sugar starts to turn brown and can eventually burn jet black. It also starts to melt together instead of remaining as separate crystals. But the salt just won't burn.

So now you know three different, simple tests for telling one chemical, sugar, apart from another, salt.

- How do they **taste**? This is a dangerous test if you don't know what you're tasting, so scientists usually skip it.
- How do they **look**? Not just with the naked eye, but up close.
- How do they **react** to heat? At a particular temperature some chemicals turn into liquids or gases, some burn, and some don't do anything but get hot.

There are a lot more tests for telling chemicals apart. Try thinking up some more. How about **weighing** a box full of each chemical? Or **smelling** them? Or **squeezing** them? Or **dissolving** them in water? Or—you get the idea. Anyway, this kind of thinking is what has gotten scientists excited about the nature of matter since we first started living on this planet.

Chemical ReAction

When is matter most exciting? When it changes right before your eyes! Practically everything that happens, from doing your laundry to igniting a rocket, involves two or more chemicals mixing together and making something new. Scientists (like you) call these "chemical reactions."

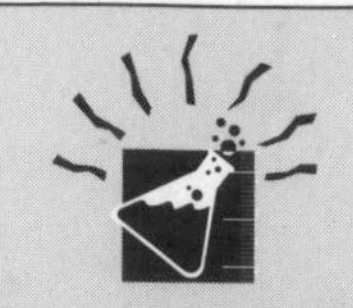

CO_2 Balloon

What You Need:

- *30 mL (2 teaspoons) baking soda*
- *60 mL (1/4 cup) vinegar*
- *A rubber balloon*
- *Straw*
- *Plastic pop bottle*

Here's What You Do:

1. *Put the baking powder into the balloon.*

2. *Pour the vinegar into the bottle.*

3. *Using your thumb, carefully put the balloon over the bottle top.*

4. *Dump the baking soda from the balloon into the bottle.*

Here's What You See:

The balloon gets bigger and bigger. The baking soda and vinegar react to produce a gas that takes up more space inside the balloon. There you are: a gas produced from a solid and liquid!

And that's not all! Feel the outside of the balloon. It's warm. The chemical reaction inside is producing **heat**, a form of **energy**. Energy can be stored or released by a chemical reaction. That's another reason why chemical reactions are so important: they not only produce new chemicals, they produce (or take away) energy. When a car burns gasoline to make the car go—that's a chemical reaction. When your body converts breakfast into energy for skateboarding—you use a lot of chemical reactions. We can't live without them.

Water Is WILD!

It's time to get deeper into chemicals. Let's start with one we all know and love—water. This experiment uses electricity to break water down into two different parts. That process is called "electrolysis" [ee-LEK-trahl-ih-sis]. It's very cool.

Let me mention that washing soda, like we will use in the next experiment, is a chemical that a scientist like you calls a **"catalyst"** *[KATT-uh-list]. It helps other chemicals react, but it doesn't get changed by the reaction. The washing soda helps the water break down, but it's actually still there.*

experiment

Electrolysis

Here's What You Need:

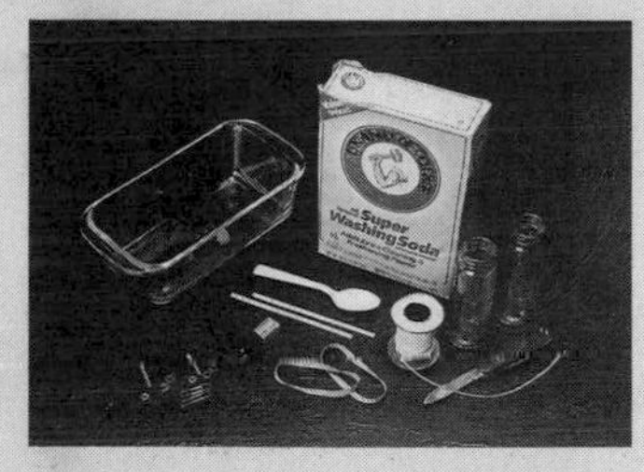

- *A 9-volt battery. The rectangular type used in transistor radios.*
- *2 identical narrow jars, like the ones olives come in, or 2 identical test tubes.*
- *2 pieces of solid insulated wire, each about 30 cm (1 foot) long.*
- *2 stainless steel screws with nuts.*
- *Deep glass baking dish, the type you make brownies in. It has to be deeper than the narrow jars are wide.*
- *Warm water*
- *Washing soda, which you can buy at the grocery store in the laundry detergent section. (This is not the same as baking soda.)*

Continued...

Here's What You Do:

1. *Strip 1 cm (1/2 in.) of insulation off the end of each wire. Strip 5 cm (2 in.) off the other end of each wire.*

2. *Wrap the long bare end of one wire around the end of one of the screws and secure it by screwing on the nut tightly.*

3. *Fill the glass baking dish with warm water. For every quart of water add about 15 mL (1 teaspoon) of washing soda.*

4. *Submerge the two olive jars in the water. Then lift up their bottoms and set them upright upside-down. Don't let any air bubbles inside the jars! If the jars' openings are wider than the water is deep, you either need more water or thinner jars. (If you use test tubes, help them to stand upright with clothespins taped to a cardboard box. Take your time to make sure they stand firm.)*

5. *Stick the two steel screws with their wires attached upward into the jars. Make sure the jars are full of warm water, make sure they'll stay up, then bend the wire for the screws so that they stay in place. Take your time to get it right. It's worth it because what you're about to do is so cool.*

6. *Connect the wires coming from the two steel screws to the battery.*

Here's what you should have after everything is all set up:

A. *Warm water with washing soda.*

B. *Two narrow jars completely filled with warm water and turned upside down.*

C. *Two steel screws with wires stripped on the ends, attached to the steel screws, and waterproofed with wax.*

D. *The two steel screws sticking up into the full jars.*

E. *The two wires attached to the electrodes of the battery.*

Here's What You See:

In a few moments bubbles form and float up off the steel screws. As they gather at the tops of the jars, the gases push out some of the water. Let the set-up sit until the jars have at least 2.5 cm (1 inch) of gas at their tops.

By now the two jars have enough gas at their tops. How much gas? Get a ruler and measure. The water in one jar should be pushed down almost exactly twice as far as the water in the other. When you break down water into two parts, you make exactly twice as much of one chemical as of the other!

Electrolysis shows that water is made of two chemicals: **hydrogen** *(H) and* **oxygen** *(O). When you divide water into these gases, you get twice as much hydrogen as oxygen.* **H_2O**, *get it? When you see the ocean, you can imagine that it is exactly 1/3 oxygen and 2/3 hydrogen. Electrolysis is used on submarines to make oxygen for the sailors inside to breathe. On a submarine, particularly a nuclear-powered one, getting a lot of electricity is not a problem. And there's plenty of water available!*

Digging Into MATTER

DEMOCRITUS

Around 450 B.C.
Conceived of atoms.

What is matter made of? You probably know the word "**atoms**" [A-tumz]. It comes from a very old Greek word meaning "uncuttable." At first atoms were just an idea that a Greek guy named Democritus came up with about 2400 years ago. He didn't have any evidence that there were atoms. He couldn't see one. He just figured that everything he could touch and feel (rocks, ice, wood, meat—matter) could be divided only so many times. He thought if you start with any type of matter, like a piece of cheese, and cut it in half and then in half again and again, it would get smaller each time you cut it. Eventually, you'd come across a piece of cheese so small that it couldn't be cut at all. That would be an atom of cheese, he thought.

Scientists didn't look into this idea of atoms for a long time because of an even more famous Greek guy named Aristotle. He wrote that everything is made from only four basic things, called "**elements**" [ELL-u-mints]: earth, water, wind, and fire. When he saw a tree, Aristotle probably thought along these lines: "A bush

ARISTOTLE

384-322 B.C.
Conceived of elements.

grows from the *earth.* It needs *water* to grow. If I seal it in a box, it will die, so it needs *air.* And if I hold a candle to it, it gives off a lot of *fire.* So a bush must be made of earth, water, air, and fire." At first that might sound a little silly, but in a way it makes sense. You have to give Aristotle credit—at least he took a shot at understanding what the world is made of.

Aristotle was such an influential thinker that for hundreds of years people believed in his four elements. Scientists called "**alchemists**" [AL-kemm-ists] tried to find the right combination of the four elements to turn things into gold. Changing stuff you had around the farm-house into gold seemed like a good way to make money. Nobody was able to do it, but in the process, the alchemists learned a lot more about matter and the science of chemistry. Our modern word "chemistry" is from the Greek words for "changing metal."

In 1774 a British scientist, Joseph Priestley, showed that oxygen is an element. It can't be separated into other chemicals. It's pure. Once scientists figured out how to look for elements, they began to discover what matter really is. It turned out Democritus was partly right about all matter being made up of little pieces. But there's no cheese atom, as he thought, because cheese is not an element. Aristotle was partly right about all matter being made from basic elements. But those elements aren't earth, water, air, and fire. And the alchemists were on the right track mixing chemicals to make other chemicals. But they were wrong about making gold. Gold is an element, so it can't be made from anything else—you'll have to find another way to get rich!

JOSEPH PRIESTLEY

1733-1804.
Proved oxygen is an element.

The Real Story of Matter

Now, 2000 years after Democritus, scientists have figured out most of the real story of matter. All the matter that we see comes in one of three ways.

Elements

The first and simplest form of matter are the elements. Each element comes in its own unique kind of **atom**. The helium in a balloon is a bunch of helium atoms floating around. A pure silver spoon is a bunch of silver atoms sticking together and formed into a bowl and a handle. Atoms come in different sizes, but they're all about a tenth of a billionth of a meter across. That's about four billionths of an inch.

$\frac{1}{10,000,000,000}$ m

COMPOUNDS

When two or more atoms stick together, they make what's called a "**molecule**" [MAHL-eh-kyool], from the Latin word for "little mass." When we have a bunch of the same kind of molecules together, we call it a "compound." Water is a compound because water molecules are made from hydrogen and oxygen atoms. Chemical reactions usually involve parts of one molecule recombining with parts of another molecule to create new molecules—like changing partners in a dance.

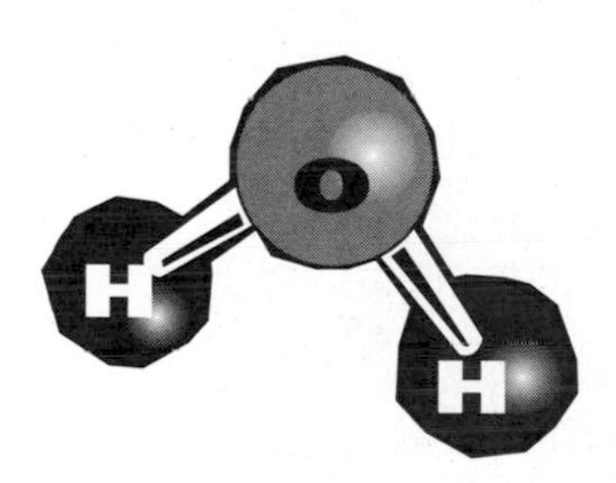

Think back to that experiment heating sugar grains over a candle flame. (If you didn't do it before, try it now!) Sugar comes in molecules made from lots of carbon atoms, hydrogen atoms, and oxygen atoms. If you heat sugar over a flame long enough, you get a completely black puddle. The puddle is almost pure carbon atoms. The hydrogen and oxygen atoms have broken off from the carbon atoms and floated off into the air.

Mixtures

Finally, some kinds of matter look like they're pure, but they're really a **mixture** of different molecules and atoms. We can separate those chemicals if we want. Air, for instance, is a mixture of gases: nitrogen, oxygen, carbon dioxide, and more. Air is not an element or a compound; it's a mixture of elements and compounds.

*One special type of mixture is called a "**solution**." Saltwater is a solution. Salt molecules are floating around in the water, each broken into two parts called "**ions**." Saltwater looks pure, but we can separate the molecules by boiling the water into vapor, leaving the salt behind. Solutions are useful because, with molecules broken apart and floating around, it's easy to make the ions fit together in new ways, making new molecules.*

SALT
H_2O

Type of Matter	Smallest Pure Part	Examples
Element	Atom	hydrogen, iron, plutonium, nitrogen
Compound	Molecule	salt, rust, carbon dioxide, soap
Mixture	not pure at all!	saltwater, milk, concrete, soap with 1/4 moisturizing cream

The Big 92

You might not expect this at first, but there are only **92 elements** in nature. (We humans have managed to make 14 more. Many of these last for only a few instants.) That means that whatever matter you come across is made of the 92 natural elements, either pure or combined into compounds or mixed into mixtures.

Look at all the different stuff around you: everything is made of those 92 elements. Rocks up in the mountains, the bones of extinct dinosaurs, the organs inside your body, the atmosphere of Mars—all made of the same 92 elements. A new medicine that hasn't even been invented yet but may save your life someday—you can be sure it will be made from some of those same 92 elements! That's highly cool.

Inside the Atom

All matter is made of molecules, which are made up of atoms of different elements. So you might wonder, what makes the elements different from one another? The answer lies in how the atoms are put together. Here's the deal:

At the center of every atom is a "**nucleus**" [NEW-klee-us]. It's actually a group of particles all hanging on to each other. There are two types of particles in the nucleus: **protons** [PRO-tahnz] and **neutrons** [NEW-trahnz]. Another type of particle is moving around the nucleus: **electrons** [ee-LEK-trahnz]. Protons, neutrons, and electrons are all called "**sub-atomic particles**," meaning "under the size of an atom."

The number of protons in the nucleus determines what element the atom is. If there are 8 protons, it's oxygen. If there are 9 protons, it's fluorine. If there are 92 protons, it's uranium. The only thing that makes the oxygen that we breathe different from chlorine, which we'd die if we breathe, is the number of protons. It's simple.

The Proton

"Proton" comes from the Greek word for "first." Protons are actually what determines what element an atom is going to be. If you talk to an atom, the first question you'd ask is, "How many protons do you have?" You won't get an answer because real atoms can't talk. But it's a good question because the answer would tell you what element the atom is. Each element has a unique number of protons called its "**atomic number**." One more important thing about protons: they have a **positive electric charge** (+).

The Neutron

A neutron is about the same size as a proton and is also found in the nucleus. But protons have a positive electric charge, and neutrons have **no charge** at all. The word "**neutron**" comes from the same Latin word as "neutral." Neutrons give atoms two important features. They determine whether an element is radioactive—more about that later. And neutrons determine how much an element weighs. Scientists call the total number of protons and neutrons in an atom its "**atomic mass**." (Electrons are so light that their weight usually doesn't have to be figured in.)

I'm a neutron...
(sigh)...no charge...

You have to deal with neutrons whenever you lift a bucket of water. A hydrogen atom has one proton. An oxygen atom has 8 protons. Therefore, all the protons in a molecule of water should add up like this: 1+8+1=10. But oxygen has an atomic mass of 16; each oxygen molecule has 8 neutrons as well as 8 protons. That means every water molecule has a mass of 1+16+1=18. That's almost twice as heavy!

Sometimes different atoms of the same element have different numbers of neutrons. For instance, there are forms of hydrogen that have one or two neutrons in their **nuclei** [NEW-klee-eye, meaning more than one nucleus]. Forms of an element that have different numbers of neutrons are called "**isotopes**" [ICE-uh-tohps].

Heavy Water

Imagine this: in front of you are two barrels of water. The containers are identical, and each contains the same amount of water. The same *exact* amount of water. The water in both barrels looks the same, tastes the same, puts out fires the same, even dissolves lemonade mix the same. But the water in the first barrel weighs 10 kilograms, and the water in the second barrel weighs 9 kilograms! How could that be?

The answer is hiding in the nuclei of the hydrogen atoms that make up the water. Start by thinking of a water molecule made with ordinary hydrogen atoms, which have an atomic mass of 1. Each water molecule made with two ordinary hydrogen atoms and one ordinary oxygen atom has this mass:

1 + 16 + 1 = 18

But what if the hydrogen atoms each contain an **extra neutron**? The atomic mass of that hydrogen isotope is 2, so the total mass of the atom is:

2 + 16 + 2 = 20

Water made with hydrogen that contains extra neutrons is heavier than ordinary water. That's why it's called "**heavy water**."

If each atom of heavy water weighs 2/18 more than an atom of ordinary water, that means any amount of pure heavy water weighs 2/18 more than the same amount of ordinary water. Think back on those two imaginary barrels. They contain the same number of water molecules. But if the first barrel contains pure heavy water, and the second barrel pure ordinary water, it makes sense for the first barrel to weigh more. Exactly 2/18, or 1/9, more! You can count those tiny neutrons just by weighing barrels. Pretty wild idea.

I'm what you might call a free electron...

The Electron

Electrons are way out away from the nucleus, and they go zipping around it—all day, all the time. Electrons have a **negative electric charge** (-), so they're attracted to the positively-charged protons. "Opposites attract."

When scientists first figured out that electrons exist, they assumed that the electrons were in orbit around the nucleus, just like the Moon is in orbit around the Earth. Later scientists discovered it's a little more complicated than that: electrons move around in strange, untraceable patterns. We know they're more likely to be in some places than others, but we never know exactly where. The electrons' patterns are called "**orbitals**" [OAR-bit-alls]. For many experiments you can still think of electrons as whirling in orbit, though.

Electrons are much smaller than protons and neutrons. It takes about 1800 electrons to weigh as much as one proton, and that proton is mighty tiny to begin with. Also, electrons are zipping around very fast. As a result, electrons quite often can be jolted off of atoms and go hang out somewhere else.

Basketball
30km (20mi.)

Inner Space

Here's one astounding fact about atoms. If an electron were the size of a basketball, the nucleus would be the size of a car. And, even weirder, the basketball would be almost 30 kilometers away. That's right, 20 miles! Think about that. It means that atoms are mostly empty space. And everything made of atoms (like you and me) is also mostly empty space. Just space. Nothing.

Static Electricity

Blow up two rubber balloons. Tie one on a string and hang it from the lamp. Rub the other balloon on your sweater, a rug, or a clean head of hair—something fuzzy. Bring it near the balloon on the string. The two balloons will push each other around.

As you rub the balloon, you're rubbing some electrons from your hair or sweater onto the balloon. You're actually moving pieces of atoms with your hand! The electrons stay on the outside of the rubber. We call them "**static**" [STA-tik] because they stay.

The electrons on one balloon push away electrons hanging out on the other balloon because they all have the same negative charge. That force can actually move the balloons around. Or you might find a spot on the tied-up balloon with fewer electrons and more protons; in that case, the protons will attract the electrons on the other balloon. They may stick together, and you can often lift the balloon on the string with the one in your hand. It's wild!

The Chemical Code

Every element is symbolized by a one- or two-letter abbreviation. **H** means hydrogen. **O** means oxygen. **C** was assigned to carbon, so copper was given **Cu** and chromium **Cr**. Sometimes the abbreviations don't come from English: gold is **Au** because the Latin word for gold is "aurum." In recent years, elements have even been named after famous scientists (einsteinium) or the places where they were discovered (californium).

Scientists refer to compounds by combining the symbols of the elements. Rather than say, "a hydrogen stuck to an oxygen stuck to another hydrogen," you can say, "H_2O." Carbon dioxide is one carbon and two ("di") oxygens: CO_2. Sulfuric acid is made from two hydrogens, one sulfur, and four oxygens: H_2SO_4. The same rules apply as compounds get more complicated and their symbols get longer: $NaC_{17}H_{35}COO$ is a type of soap.

The Elements

Element	Symbol	Atomic Number	Atomic Weight
Actinium*	Ac	89	227
Aluminum	Al	13	26.98
Americium**	Am	95	243
Antimony	Sb	51	121.75
Argon	Ar	18	39.95
Arsenic	As	33	74.92
Astatine*	At	85	210
Barium	Ba	56	137.34
Berkelium**	Bk	97	245
Beryllium	Be	4	9.01
Bismuth	Bi	83	208.98
Boron	B	5	10.81
Bromine	Br	35	79.90
Cadmium	Cd	48	112.40
Calcium	Ca	20	40.08
Californium**	Cf	98	248
Carbon	C	6	12.01
Cerium	Ce	58	140.12
Cesium	Cs	55	132.91
Chlorine	Cl	17	35.45
Chromium	Cr	24	52
Cobalt	Co	27	58.93
Copper	Cu	29	63.55
Curium**	Cm	96	245
Dysprosium	Dy	66	162.50
Einsteinium**	Es	99	253
Erbium	Er	68	167.26
Europium	Eu	63	151.96
Fermium**	Fm	100	254
Fluorine	F	9	19
Francium*	Fr	87	223
Gadolinium	Gd	64	157.25
Gallium	Ga	31	69.72
Germanium	Ge	32	72.59
Gold	Au	79	196.97
Hafnium	Hf	72	178.49
Helium	He	2	4.00
Holmium	Ho	67	164.93
Hydrogen	H	1	1
Indium	In	49	114.82
Iodine	I	53	126.90
Iridium	Ir	77	192.22
Iron	Fe	26	55.85
Krypton	Kr	36	83.80
Lanthanum	La	57	138.91
Lawrencium**	Lr	103	257
Lead	Pb	82	207.2
Lithium	Li	3	6.94
Lutetium	Lu	71	174.97
Magnesium	Mg	12	24.31
Manganese	Mn	25	54.94
Mendelevium**	Md	101	256
Mercury	Hg	80	200.59
Molybdenum	Mo	42	95.94
Neodymium*	Nd	60	144.24
Neon	Ne	10	20.18
Neptunium**	Np	93	237.05
Nickel	Ni	28	58.71
Niobium	Nb	41	92.91
Nitrogen	N	7	14.01
Nobelium**	No	102	253
Osmium	Os	76	190.2
Oxygen	O	8	15.99
Palladium	Pd	46	106.4
Phosphorus	P	15	30.97
Platinum	Pt	78	195.09
Plutonium**	Pu	94	244
Polonium*	Po	84	209
Potassium	K	19	39.10
Praseodymium	Pr	59	140.91
Promethium*	Pm	61	145
Protactinium*	Pa	91	231.04
Radium*	Ra	88	226.03
Radon*	Rn	86	222
Rhenium*	Re	75	186.2
Rhodium	Rh	45	102.91
Rubidium*	Rb	37	85.47
Ruthenium	Ru	44	101.07
Samarium*	Sm	62	150.4
Scandium	Sc	21	44.96
Selenium	Se	34	78.96
Silicon	Si	14	28.09
Silver	Ag	47	107.87
Sodium	Na	11	22.99
Strontium	Sr	38	87.62
Sulfur	S	16	32.06
Tantalum	Ta	73	180.95
Technetium*	Tc	43	98.91
Tellurium	Te	52	127.60
Terbium	Tb	65	158.93
Thallium	Tl	81	204.37
Thorium*	Th	90	232.04
Thulium	Tm	69	168.93
Tin	Sn	50	118.69
Titanium	Ti	22	47.90
Tungsten	W	74	183.85
Unnilquadium*†	—	104	257
Unnilquintium*†	—	105	260
Unnilsextium*†	—	106	263
Uranium*	U	92	238.03
Vanadium	V	23	50.94
Xenon	Xe	54	131.30
Ytterbium	Yb	70	173.04
Yttrium	Y	39	88.91
Zinc	Zn	30	65.37
Zirconium	Zr	40	91.2

**** Radioactive commonly***
***** Radioactive and human-made***
† Temporary name

Some *Special* Elements

Radioactive Elements

There are a few elements that are what we call "**radioactive**." Parts of their nuclei are beaming out of their atoms all the time. ("Radio" in Latin means to "beam," or go out in a ray.) How? The nuclei **decay**. They fall apart. In other words, radioactive elements are constantly disintegrating.

Uranium is one radioactive element. If you leave a piece of uranium alone long enough, it turns into a totally different element, lead. This is called "**transmuting**" [tranz-MYOOT-ing]. Lead isn't radioactive; it doesn't transmute. That's why lead is used to shield people from radioactivity.

You might wonder, "In that case, how is there any uranium around at all? How come it hasn't *all* turned to lead?" It takes a long time for uranium to decay, about 70 billion years. The Earth was formed about 4 billion years ago, so there's still plenty around for us now.

Human-made Elements

We humans have also found ways to make 14 new elements that don't exist in nature. They showed up only in the last fifty years after nuclear explosions and experiments with radioactive elements. Some last only a few instants—just long enough for scientists to measure that they exist—before transmuting into other elements.

You may have heard of the human-made element called **plutonium**, made by beaming extra neutrons into uranium. It is *extremely* radioactive. And it lasts a very long time. The guy who named Plutonium, Glenn Seaborg, assigned it the symbol "**Pu**" to show other scientists that it's very dangerous, in a way, and "bad smelling" stuff.

On Beyond Atoms

Aristotle thought that water was an element, that it couldn't be divided into any particles that were not water. Then Priestley showed that a molecule of water could be divided into atoms of hydrogen and oxygen. Then scientists discovered that each atom could be divided into protons, neutrons, and electrons. What's next?

and Energy Matter

By using very large machines that make protons or neutrons spin very fast and collide, we've figured out that there are a few more particles even smaller than electrons. But we still haven't figured out exactly how these particles work. One reason it's hard to get a handle on the tiniest of particles is because when things are this small, they don't act like the matter we regularly see. For one thing, these particles of matter frequently turn into energy!

You've probably heard of Albert Einstein's famous equation: $\mathbf{E = mc^2}$. Those aren't chemical symbols. Instead, **E** is the symbol for energy, and **m** is for mass, which means the amount of matter. And **c** means the speed of light, which is 300,000 meters per second. The little **2** means the speed of light *times* the speed of light: 90,000,000,000 watt-seconds! What all that means is that just a little mass (m) can make a lot of energy (E). If you could convert one gram of water completely into energy, you'd get enough energy to shoot a baseball from here to the Sun in four seconds. We'd have a huge amount of energy in every bucket of water! Unfortunately, converting mass into energy is hard to do with big things. But it happens all the time in the world of sub-atomic particles—pretty weird.

$$E=MC^2$$

ALBERT EINSTEIN

The Positron

There's a particle as tiny as an electron that has a **positive charge** like a proton. It's called a "positron" [PAH-zih-trahn] or "anti-electron." That's right: positrons are what's called "**anti-matter**," just like in science fiction. The best place to find positrons is outer space. Now get this: if a positron meets an electron, they destroy each other and become the energy form called a "**gamma ray**." That's like an x-ray. It's another case of what's really going on in science is way cooler than anything we make up.

Quarks and Gluons

Near as we can tell, protons and neutrons are really made of even smaller particles called, of all things, "quarks" [kwarks]. Now, we've never actually found a quark, but it sure looks like we should be able to. As the theory stands right now, each proton should be made of two **Up-quarks** and one **Down-quark**. And, each neutron should be made of two Down-quarks and one Up-quark. The quarks are imagined to be held together with gluons [GLUE-ahnz], or "glue particles." We'll see. In the next few years, scientists should be able to figure this out.

Neutrinos

The quarks and gluons, in a way, are handing energy back and forth to each other. When they collide, sometimes they convert the quarks of neutrons into the quarks of protons—or the other way around. And, in the process, sometimes they make an electron and they give off an even *even* smaller particle called a "neutrino" [new-TREEN-oh]. Neutrinos, by the way, seem to be everywhere in the universe. Near as we can tell, they're zipping right through Earth and through you and me all the time.

So we can make a list of things from **BIG** to **small**:

	Object	Distance Across
BIG	The Moon	3480 km (2160 mi.)
	Hot Air Balloon	20 m (60 ft.)
	Basketballs	25 cm (10.5 in.)
	Orange	10 cm (2.5 in.)
	Dust Particle	10 microm. (0.0005 in.)
	Molecule	10 nm (0.4 millionths of an inch)
	Atoms	0.1 nm (4 billionths of an inch)
	Neutrons & Protons	0.00001 nm
	Electrons & Positrons	1000th of a neutron
	Quarks	no one is sure
small	Neutrinos	ten billionth of a neutron

And beyond neutrinos? We don't know. We're not even sure about quarks. But figuring that stuff out is what makes science exciting. Whatever the answer turns out to be, it will fit in with what we know so far. It has to. Furthermore, most scientists believe that however an atom is really constructed, it won't be too complicated. If there are only five or six particles that make up everything, and we can understand those, we might understand every thing in the whole universe. Whoa! Here's hoping you'll be the one to figure it out.

WHOA!

THE FOUR FUNDAMENTAL FORCES

chapter three

The universe wouldn't be very exciting if everything stayed the same. For things to change, matter has to move. Sometimes matter moves in ways we can see: when we wiggle our ears, for instance. Sometimes the matter that moves is too small for us to see: when atoms come together to form molecules. But either way, something has to make that matter move.

Off the top of your head, you can probably name hundreds of changes in the universe. Food cooks in an oven. The Sun heats the air. Leaves change color. Birds fly. Plants grow bigger. The Earth quakes once in a while. The planets revolve around the Sun. And you change when you grow taller, or get a cold for a few days, or fall asleep at night. All those changes depend on matter moving. And, near as we can tell, all movements of matter are caused by just four forces—what we scientists call the "**Four Fundamental Forces of Nature**."

Gravity

As you read these words, you and this book and whatever you're sitting on are being pulled down by **gravity**. If you're reading in an airplane seat, the plane is being pulled down by gravity. That's why planes need wings and engines. Only if you're reading in outer space (and that's not likely yet) do you not feel the force of gravity—but it's still there. It's a **Fundamental Force**. We feel gravity almost all the time, but it's most obvious when things (or people) fall down.

The Error from Aristotle's Era

Aristotle believed that heavy things (and, presumably, people) fall faster than light things. He satisfied himself that this was true by dropping a rock and a feather. No one questioned his theory for several centuries, even though it's easy to see that his idea just doesn't fly.

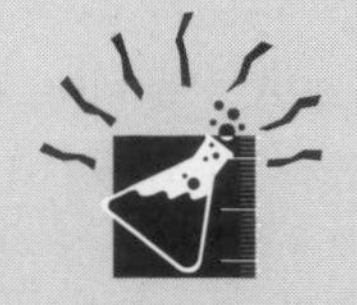

Double Drop

Take a sheet of paper and a ball. Hold them up in the air at the same height. Drop them. Aristotle's hypothesis predicts that the ball lands on the floor before the paper because the ball is heavier. And sure enough, the ball does land first.

But Aristotle never tried this part of the experiment. Crumple the paper into a wad, as small as you can make it. Hold the wad and the ball at the same height, and drop them. Now the ball and the paper land at the same time. The paper hasn't gotten any heavier; it's just in a different shape. A round wad of paper can push air out of the way more quickly than a flat sheet of paper. If there was no air to push out of the way, even the sheet of paper would plummet at the same speed as the ball.

Galileo's Gravity

One of the first people to understand gravity was Galileo Galilei, an Italian scientist. We usually just call him "Galileo" [GAL-ih-LAY-oh]. He did many experiments with gravity and took careful notes of what he saw. He understood that the Earth pulls objects towards its center—the direction we call "down." Galileo also understood that it doesn't matter how big objects are, they all fall down at the same rate.

It is said (this may or may not be true) that Galileo went to the Leaning Tower of Pisa, stood at the very top of this very tall building, and dropped two rocks or cannonballs of unequal size over the edge. They fell about 55 m (180 feet) to the ground at precisely the same speed, so they landed at precisely the same time. I like to imagine Galileo shouting down to passersby, "Hey, did you see that? Things fall toward the

GALILEO GALILEI

1546-1642.
Showed gravity is constant.

center of the Earth at the same speed!" Even if the Leaning Tower of Pisa story isn't true, it's cool. Starting to understand how gravity really works was a turning point for science.

ISAAC NEWTON

1642-1727. Discovered rules of gravity.

Galileo realized that gravity does more than make things fall towards the center of the Earth. He also figured out that the Earth's gravity holds the Moon in orbit around it. Then he realized that the Sun has its own gravity which holds the planets, including Earth, in orbit around it. Based on Galileo's work, the English scientist Isaac Newton worked out the exact rules of gravity. Here they are in simple form:

GRAVITY RULES:

All matter has gravity. Gravity isn't just for big things like the Earth. Even you have gravity. Even an atom has gravity.

Gravity pulls, or attracts. It never pushes, or repels. Every piece of matter in the universe is pulling on every other, at least a little.

The more mass, the more gravity. A large mass like the Earth pulls more than a smaller mass like the Moon. That's why astronauts can bounce around on the Moon, which has less gravitational pull. Does that mean that big, heavy people have more gravity than small, light people? In fact, they do. But since the Earth is more than 100 septillion times larger than the average adult, people's gravity is extremely tiny compared to the gravitational pull of the Earth. We can't feel each other's gravity, and we certainly don't stick to big people as we walk down the street.

The farther away, the less gravity. As you move farther away from an object, even a tremendously massive one, its gravity pulls on you much less. That's why astronauts feel weightless when they're in space far from the Earth or the Moon. It's wrong to say there's no gravity in space. Gravity is still working throughout the universe, but the astronauts aren't near enough to a huge object to feel its gravitational pull.

Right now, nobody is exactly sure why matter has gravity. But we do understand exactly how it behaves, and we think it behaves the same everywhere. So scientists (like you) call gravity a Fundamental Force.

Electromagnetism

Think back on atoms, as described in Chapter 2: electrons moving around the nucleus of protons and neutrons. What holds the electrons there? Why don't they just go sailing off into space? They stay because they're attracted by the opposite electric charge of the protons. That's the basis of electromagnetism [ee-LEK-trow-MAG-neh-tizz-um], another Fundamental Force of Nature.

Electromagnetism works on every piece of matter that has an electric charge, not just electrons and protons. It causes atoms to stick together in molecules, producing every kind of chemical reaction. And you probably watch electromagnetism work every day with magnets and refrigerator doors. That's all part of one Fundamental Force.

To read this book you're using electromagnetism in several ways:

- *Light bounces off the book into your eyes. Chapter 5 explains how light is a form of electromagnetic energy. As they say on television, "Stay tuned!"*

- *The ink sticks to the pages of this book because of electromagnetic attraction between the ink molecules and the paper molecules.*

- *When you think about what you're reading, chemical reactions happen inside your brain. Each of those reactions depends on the electromagnetic attraction of electrons and protons.*

- *You even need electromagnetism to hold this book. Atoms are mostly empty space—almost unimaginably empty. But electrons on the outside of the book's atoms and electrons on the outside of your atoms have the same electromagnetic charge, so they push each other apart. The book repels your fingers, which makes it feel solid. So strong is the repelling force of electrons that it lets us feel matter as solid, or wet, or windy.*

e x p e r i m e n t

Fighting Forces

Place a steel table knife on the ground. Put a refrigerator magnet on top of it. Now the knife is being pulled by both the Earth's gravity and the electromagnetic attraction of the magnet. Which force is stronger? Lift the magnet up to see. Unless you've chosen a very heavy knife and a very small magnet, the magnet will lift the knife right off the ground. Think about that. A tiny magnet, smaller than your little finger, overcomes the gravitational pull of the entire planet! **Electromagnetism** is about a **trillion trillion trillion** times **stronger** than **gravity**.

1,000,000,000,000,000,000,000,000,000,000,000,000

Forces Strong and Weak

In atoms, the protons all have a positive charge, and they're all squeezed together in the nucleus. But if electromagnetism means that positively charged particles push each other away, why don't the protons repel each other and go flying off in opposite directions? Because an atomic nucleus is held together by something we call the "**strong atomic force**." As you may have guessed, that's another Fundamental Force of Nature.

Without the strong atomic force, all the positively charged protons would repel each other like crazy. They would never make a nucleus, which means no atoms, which means no elements, and so on. The universe would be a big cloud of banging, bumping, unorganized particles. There would be nothing recognizable to us. There would be no "us," if you see what I mean.

The strong atomic force is very strong indeed. It's a thousand times stronger than the electromagnetic force. But it seems to work only over very very short distances. I mean distances so short we can barely imagine them: about a quadrillionth of a meter. That's 1/100,000 of the distance across an atom. Small small small.

From Chapter 2 you might also remember that some elements can be **radioactive**. Certain types of atoms have pieces shooting out of them all the time. That's **nuclear radiation**. It's how we make nuclear power and atom bombs. In these atoms, the strong atomic force isn't strong enough to hold everything together. These atoms' nuclei change because of the fourth Fundamental Force of Nature, called the "**weak atomic force**." The weak atomic force creates radioactivity. Like the strong atomic force, it only works across a very small distance deep inside atoms.

The Four Fundamental Forces (from strongest to weakest)

FORCE	RANGE	WHAT IT DOES
Strong Atomic Force	.000000000000001 m	Holds protons and neutrons in the nucleus
Electromagnetism	Infinite	Holds atoms and molecules together; pulls particles of opposite charges together; pushes same charges apart
Weak Atomic Force	.0000000000000001 m	Allows atomic radiation
Gravity	Infinite	Pulls all types of matter together; holds us on Earth

TOE?

Feel It In Your GUT

GUT?

There are only four Fundamental Forces of Nature, as far as we know. (If more exist, they are probably hidden deep inside atoms.) Apparently that's the way the universe is set up. Many scientists are looking for ways to combine all the theories of these Four Forces into one theory to describe them all at once. They're looking for a **Grand Unified Theory** (GUT) or a **Theory of Everything** (TOE). Maybe you'll be the one to figure out how to make it all fit together. If you do, you might know how the whole universe works.

WOW!

chapter four

HEAT

What does heat feel like? Like a warm blanket, a fever, a cup of cocoa, a scorching hot beach. Ouch! But what is heat? It's energy. Pure energy! And energy, together with matter, is what makes our universe such a happening place. Heat is the kind of energy that makes atoms and molecules vibrate.

When you pluck a guitar string, you're making it vibrate. If you've got a guitar, strum it and look at one string closely. Or pluck a rubber band. You can see it move. It's moving up and down very fast, but it's staying in pretty much the same place. That's what molecules and atoms do when they are heated—when heat energy has been put into them. They **vibrate**, moving up and down and back and forth in one place very fast. But remember, these are atoms, and they're tiny. Actually, they're way past tiny. So when they vibrate, we can't see it. But we can notice their vibration as heat.

All Steamed Up

What happens when you put more heat energy into water? Well, the water molecules start to vibrate. And if they vibrate enough, the water changes.

The Vapor Balloon

Here's What You Need:

- *Water*
- *Pencil*
- *Rubber balloon*
- *Microwave oven*
- *2 hardcover books*
- *Sheet of paper*

Here's What You Do:

1. *Put 200 mL (1/2 cup) water in a balloon.*
2. *Fill the rest of the balloon with air so it's about 10 cm (4 inches) wide, about as big as a large grapefruit. Sometimes water squirts back into your mouth as you blow it the rest of the way up, so be ready.*
3. *Measure the balloon. You can do this by placing it on top of a piece of paper and just barely squeezing it with two upright books. Mark the paper where the books were, and you'll have a pretty good way of knowing how big the balloon is.*
4. *Put the balloon in a microwave oven. Never* **WARNING!** *turn on a microwave oven with no water in it. Most microwave ovens can be ruined by running them without 200 mL (1/2 cup) of water inside to absorb energy. Luckily, food usually has a little water in it.*
5. *With water in the balloon, turn the microwave oven on for about a minute and a half.*
6. *Take the balloon out and measure it again with your books and the same piece of paper. Has it changed?*

Here's What You See:

The balloon sits in the microwave for while, and then it just starts to get bigger. You can see it. And your measurement confirms that it grows. What happens inside the balloon? The molecules of air and water vapor got hot and took up more room in the balloon—a lot more. What do you think moves them apart? It's **heat**. The energy of heat makes the molecules of water vibrate. If we get them to vibrate

enough, they go flying off as a **gas**. And they remain in the gas phase as long as they have enough energy to keep vibrating fast—as long as they have enough heat energy in them.

Think of it this way: Imagine you're in gym class, and everybody's lined up for the teacher to take attendance. Then the teacher tells everyone to warm up (it's always smart to warm up before exercising) with a few jumping jacks. What do you students do? You spread out, of course. Because as soon as you start jumping, you start bumping each other every so often. That's sort of what happens with molecules when they're heated: they start **vibrating**, so they spread out.

Now try taking heat away from the water balloon. Take the warm balloon and put it in the freezer. It takes longer for the balloon to get cold in the freezer than it does for it to get hot in the microwave. Take the cold water balloon out after 5 or 10 minutes and measure it on the paper. It shrinks to a much smaller size than what you started with. See, the molecules of water vapor that were already floating around in the balloon when you started are going back to being a **liquid** in the freezer, and they take up a lot less space. It's cool . . . er, cold.

Finally, just to drive this idea home, set the cold balloon down on a counter top and let it come back to room temperature. Measure it again. It'll be the same size as when you started.

In the water balloon experiment, we measured how much room a certain amount of matter took up under different temperatures. We didn't add or take away any matter. We just changed how much the molecules vibrated by changing how much heat they stored. And that determines how much the molecules spread out or pull together.

When something is heated, it's molecules get farther apart and it gets bigger, we say it **expands**. *It's from the Latin word for "spreading out." When something is cooled, its molecules move closer together and it gets smaller. We say it* **contracts** *[kon-TRAKTS], from the Latin word meaning to "draw together." All objects expand and contract when they're in any phase—solid, liquid, or gas.*

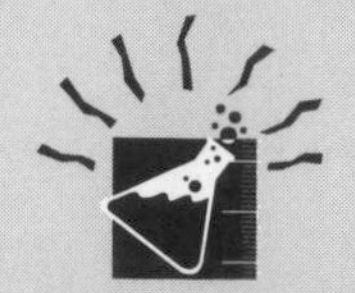

Expanding and Contracting

Here's What You Need:

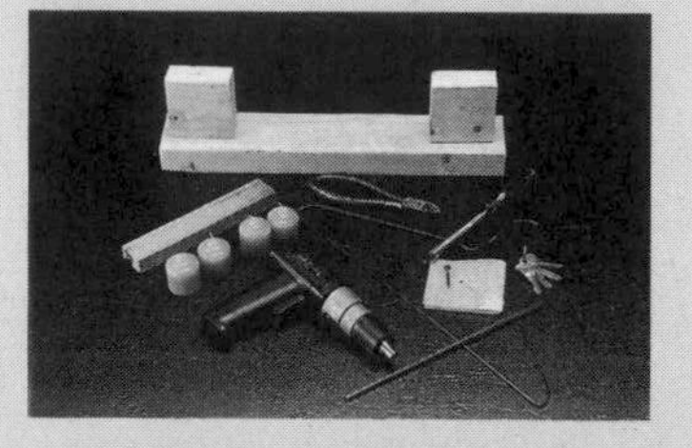

- *A nice straight piece of wire coat hanger. You may need to have an adult cut you a piece of an old hanger with wire cutters.*
- *Piece of wood*
- *Screw*
- *Screwdriver*
- *Drill*
- *Pin. A hat pin works well, or just a large thick straight pin.*
- *Weight, such as keys*
- *Straw*
- *2 or 3 lighted candles, although only one can do.*

Here's What You Do:

1. *Place one end of the metal wire into a hole in a piece of wood and screw that end down tight.*

2. *Set the other end of the wire on top of a pin.*

3. *Weight the far end of the wire so that it's putting a little pressure on the pin—so it makes good contact.*

4. *Put a straw on the end of the pin, so with the metal resting on the pin, the straw will move if the pin is rotated, even just a little bit.*

5. *Heat the metal with a candle. If you have two or three candles, it's even better because you can warm the metal more quickly. It won't cool off at one end while the middle is getting hot.*

Here's What You See:

The metal expands, the pin rolls a little under the wire, and the straw moves as a pointer. See, it's not moving very much, but it is moving. Almost all materials expand when they're heated, some a lot more than others. That's the important scientific principle called "**thermal expansion**." The word "thermal" comes from the Greek word for "heat."

Next time you're walking over a long bridge, look for the **expansion joints**. *They're usually interlocking fingers of metal, like big combs. Sometimes they're covered up with a rubber strip. Expansion joints allow the very large pieces of concrete and steel that form the bridge to expand and contract with the air temperature. You might not think that something as solid as a bridge actually moves around just because the weather is changing. Some very large bridges have joints that allow them to expand over a meter—3 feet!*

Thermometers

The device you built in the last experiment is a thermometer. That means "something that measures heat." If you keep careful track of the position of the straw, you'll see it moves right along with the temperature of the air around it. When the room is cold, the wire contracts a little, and the straw turns counter clockwise. When the room gets warm, the straw moves slightly clockwise. Of course, you have to be careful not to disturb your set up.

Thermometers depend on thermal expansion to work. In fact, most thermometers that have a dial work with metal strips that expand and contract with temperature. Their wires are wound in a coil to take up less room, but the metals expand and contract with temperature just as the coat hanger wire did in the experiment.

Other thermometers have a long glass tube with liquid inside. Usually the liquid is mercury, a silvery liquid metal, or alcohol, dyed red. Both liquids expand evenly when it gets warm, and contract when it gets cool. That makes the top of the liquid go up and down in the tube.

We can also hook up a thermometer made with a metal coil to an electric switch. This set-up is called a "**thermostat**" [THERM-oh-stat]. (In Latin "stat" means "stay.") So a thermostat makes your house stay at the temperature you set it at. When it gets too cold, the metal contracts and trips the switch. Then the heat comes on until the house is so warm that the metal expands again and turns the switch off.

How come we invented thermometers? After all, temperature is something our bodies sense, just like smell, hearing, seeing, and taste. But we need thermometers because we want to measure temperatures of things that are too hot or cold to handle, and because it's easy for us to fool ourselves.

e x p e r i m e n t

Temperature Trick

Get three bowls big enough to put your hands in. Fill one of them with very warm (but not boiling hot) water, and fill the second one with cold water. Then pour equal amounts of hot and cold water into the third bowl.

Okay now, put one hand in the warm water and the other hand in the cold water for, say, a minute. Now one at a time put your hands in the in-between bowl of water. The hand in hot water will sense cold, and the hand in cold water will feel warmth. It's hot and cold at the same time!

Actually, the water in the third bowl is not two temperatures. It's just one temperature, somewhere between very warm and cold. But it feels different to each hand. That happens every time. Our sense of temperature depends on where our body's been. We do our best, but we're not that good at sensing temperature all the time. That's why we invented thermometers. Once you have a thermometer, you don't need to touch things to see how warm or cold they are. This can be mighty handy (no hilarious pun intended).

Temperature Scales

So how do we measure temperature? In the U.S. we have two systems: **Fahrenheit** [FAIR-enn-hite] and **Celsius** [SELL-see-us]. You might wonder, hey, what's the difference? Well, this is indeed a very good question. Fahrenheit and Celsius are to measure the same amounts of heat, but they use different sets of numbers to do it. These numbers are called the thermometer's "**scale.**" (It's from the Latin word for "ladder." The markings on glass tube thermometers do look a little bit like a ladder.)

Okay now, imagine that you have a glass tube full of alcohol with ladder-like markings on the side. What numbers are you going to write there for your scale? Are you going to start with one and go to 10? Would you start with 7 and go to 538? Or what? Hmmmm. While you're at it, *where* should you start writing the numbers? The middle? The top? The bottom? Double hmmmm.

A German scientist named Gabriel Fahrenheit, around the year 1714, had a great idea. Fahrenheit realized that the element mercury was just about perfect to use in a thermometer. See, mercury is a liquid at regular temperatures, yet it's a metal. It expands and contracts a lot with temperature, but it doesn't evaporate inside the tube. For years and years mercury has been the best thing for thermometers.

After he got the thermometer working, Fahrenheit set up his own scale for measuring temperature. He was measuring the temperature of chemicals in water, so he wanted a scale that would go *past* water's freezing point to some other cold temperature. He found that when he added a special type of salt to water and ice, it lowered the freezing temperature of the mixture. And later when

GABRIEL FAHRENHEIT

1686-1736. Invented mercury thermometer.

he mixed the water, ice, and salt the same way, it would always come out to the same coldness. Fahrenheit figured this was about as cold as anything that he was measuring would ever get, so he called it zero.

Now, Fahrenheit, like a lot of people, thought the number sixty was great. (The ancient Babylonians started this sixties business. In particular, they liked the numbers 60, 180 and 360, because these numbers can be divided by so many smaller numbers. That's why our minute has 60 seconds. That's why there are 360 degrees in a circle.) So Fahrenheit took the temperature where pure water boils and said this will be 180 degrees above where pure water freezes. Well, with all that, water's freezing point ended up at 32 degrees Fahrenheit, which we abbreviate as 32°F, and its boiling point wound up at 212°F. That's kind of a weird story, but it's kind of a weird scale for temperature.

We scientists actually use the Celsius or "**centigrade**" [SENT-ih-GRADE] scale. It's named after a Swedish scientist named Anders Celsius. In 1742 he suggested to other scientists that it would be easy to work with temperatures if we just say water freezes at zero and boils at 100. We abbreviate those temperatures as 0°C and 100°C. This scale, for a long time, was called the "centigrade" scale. ("Cent" is Latin for 100, and "grade" is Latin or French for "degree" or "step." So centigrade is 100 steps. Easy.) Then in the 1960s an international group of scientists decided to rename these degrees after the guy who invented them. So "degrees centigrade" is now called "degrees Celsius."

Water always freezes at the same temperature, no matter whether we call it 0°C or 32°F. That's the most important thing to remember about temperature scales:
The same things happen at the same temperatures, no matter what they're called.

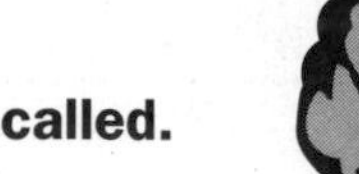

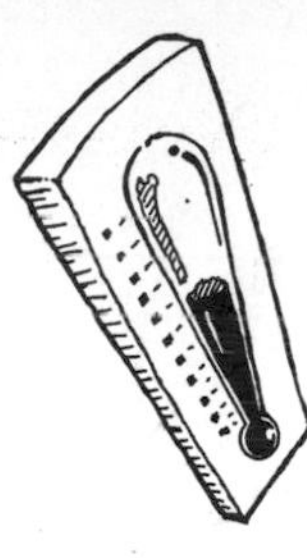

Handy Temperatures to Know:	Fahrenheit	Celsius
absolute zero	-459.67°F	-273.15°C
nitrogen turns to liquid	-320.4°F	-195.8°C
average on the Moon (night)	-292°F	-180°C
lowest ever recorded on Earth	-127°F	-88°C
carbon dioxide turns to solid	-109.5°F	-78.6°C
water turns to solid (freezes)	32°F	0°C
average in Kansas City	63°F	17°C
baby bath water	95°F	35°C
average body temperature	98.6°F	37°C
adult bath water	104°F	40°C
highest ever recorded on Earth	136°F	58°C
safe for frying hamburgers	155°F	68°C
average on the Moon (day)	266°F	130°C
water turns to gas (boils)	212°F	100°C
forest fire	570°F	300°C
steel turns to liquid (melts)	2700°F	1500°C
surface of the Sun	10000°F	5500°C

To change from Fahrenheit degrees to Celsius degrees:

1. subtract 32
2. multiply by 5
3. divide by 9

To change from Celsius degrees to Fahrenheit degrees:

1. multiply by 9
2. divide by 5
3. add 32

To get the right answers, be sure to follow these steps in the right order. And keep track of minus signs!

Measuring Heat

As a scientist you might look at a cup of hot cocoa and wonder, "How much heat is in there?" Or you might ask yourself, "What temperature is that cocoa?" Or, "How hot is it?" And at first, they all sound like the same question. But the amount of **heat** in something is not quite the same as that thing's **temperature.**

Suppose you had a bucket of hot cocoa, and your fellow scientist has that cup of hot cocoa. You two measure the temperature of each container. The temperatures are the same. Does that mean the bucket and the cup contain the same amount of heat? Well, because there's more cocoa in the bucket, it contains more **heat energy**. That makes sense. It has enough energy to get a whole bucket of molecules up to a certain temperature, while the cup has enough energy to heat a smaller amount of cocoa to that temperature.

Now here's a weird thing to think about: The bucket of cocoa could actually have a lower temperature than the cup of cocoa and still contain more energy. That's because there are so many more molecules vibrating in the bucket that, even if they vibrate more slowly than the molecules in the cup, their total energy adds up to more. Cold cocoa with more energy than hot cocoa—strange at first, but it makes sense.

The *Laws* of Thermodynamics

Heat is very important to the workings of everything in the universe. Our bodies give off heat from the many chemical reactions that are happening inside us all the time. Cars run by burning fuel. Trees use heat from the Sun to manufacture food for themselves. Almost every moving thing runs on heat, and scientists sometimes call these things "**heat engines**"—car engines, airplanes, electrical engines, and so on. The study of the what happens as heat moves around is called "thermodynamics" [THER-moh-DYE-namm-icks]. It's Latin for the "flow of heat."

The big ideas in heat are called the Laws of Thermodynamics. Now, we can't prove them, but as near as we can tell, they're absolutely true. That's why we call them laws. So important and far-reaching are the laws of thermodynamics that some people try to use them to explain what governments should do, why people fall in love, and why there are car wrecks. But we can probably depend on them only when we're dealing with heat and engines. So here goes.

The Zeroth *Law* of Thermodynamics

If you leave a bowl of hot water out for a few hours, it loses heat to the air. Eventually, the water is the same temperature as the air. The water and air reach what scientists call "**thermal equilibrium**" [ee-kwih-LIB-ree-um]. It's from the same Latin word as "equals." So the heat situation is equal. No more heat flows from the water to the air. Now this seems obvious, doesn't it? Things at the same temperature are, uh. . . at the same temperature. So that's the most basic Law of Thermodynamics.

If two things are at the same temperature, no heat will flow between them. If they are next to each other, they will be in thermal equilibrium.

This idea that things at the same temperature are in thermal equilibrium is so basic that scientists call it the "Zeroth Law of Thermodynamics." It's the law that we need even before we have a First Law! The "Zeroth"—it's a simple idea that's really important.

The First *Law* of Thermodynamics

Now let's talk about one of the most important ideas in all of science. Many scientists consider the First Law of Thermodynamics to be the most important scientific law there is. It's Number One!

Energy is always conserved throughout the universe.

To be "conserved" means that it always, always adds up. Energy can never be created or destroyed. You never end up with more energy than you started with, and you never end up with less. Heat is one form of energy, but there are other forms as well: the energy of something moving, the energy stored in a fuel, and so on. But if you add up all that energy, it always comes out the same.

Life is a Roller Coaster?

Okay fine, what does "**energy is always conserved**" mean? Let's take an exciting example—a roller coaster. Picture if you will a roller coaster going up to the top of its biggest hump. We have to use some energy to get it up there. Most roller coasters use electric motors—electric energy.

When we get to the top of the hill, we can think of our roller coaster as storing the energy we put into getting it up there. We now can go for a wild ride—down the huge big hill, around some turns, maybe upside down for a few moments, and eventually we end up at the bottom of the track. Now notice this: We never can get our roller coaster up a hill higher than the one wc started with. The energy of the roller coaster never gets bigger once we let it go and ride down. To get up higher than the first hill, we would have to add more energy from the electric motor, or by getting out to push. The roller coaster's energy remains the same. We say the energy is "conserved." That's the First Law of Thermodynamics.

Conserving Energy

Here's What You Need:

- *Baseball*
- *Ball of string*
- *High horizontal bar. A swing set on a playground works great.*
- *Scissors*
- *Tape*

Here's What You Do:

1. *Tie one end of the string tightly around the baseball several times. Tape over the string. Make sure the ball can't fall out.*
2. *Hold onto the ball of string, and throw the baseball over the horizontal bar.*
3. *Catch the swinging baseball, tie the string off so that the baseball hangs right around your stomach, and cut the string loose from the ball. Make sure the ball is hung so that it won't come off!*
4. *Now this is cool. Hold the baseball up to your nose and back away. When the string is straight, stand perfectly still and let go of the ball.*

DON'T PUSH ON THE BALL! Just let go gently.

DON'T MOVE FORWARD! If you step forward, you're in trouble! But if you hold your ground, you're in for a wild sensation.

Here's What You See:

The ball will swing away from you, swing back, and just miss your nose! It looks like the baseball is about to smack you right in the face; but it doesn't. Phew!

What saved your face from being smacked with the baseball? The swinging ball has only as much energy as you put in originally. It can't go any higher than it starts, because energy is always conserved. That's the **First Law of Thermodynamics** in action.

HOLD IT! HOLD IT!

We were just talking about roller coasters and then a ball on a string. What do these things have to with the flow of heat—with thermodynamics? Ah ha, it has a lot to do with it. Where does the heat come in? Well, you probably notice that every time you the ball swings, it misses your face by more and more. It seems to *lose* energy with each swing. Well, if all energy is conserved, where does the ball's energy go?

In 1846 a British scientist named Robert Joule wondered about the very same sort of question. He put together a machine to keep track of all the energy he put into lifting a small weight. He attached the weight to a string so that when the weight fell back down, the string turned a small paddle wheel in a container of water. Then he let go of the weight. It fell. The paddle wheel turned. Then everything stopped. Everything looked like it did before Joule lifted the weight—where was his energy now? Joule carefully measured the heat of the water and found that it had risen by *exactly* the same amount of energy as he had put into the system by lifting the weight. The energy of the falling weight had turned into **heat energy**.

ROBERT JOULE

1818-1889. Invented a way to measure energy transfer.

So where does the energy of the swinging baseball go? It becomes heat. That's right. The air pushed out of the way by the weight is being swirled around and heats up—just a little, but enough. And the string, which is being stretched with each swing, is ever so slightly heating up. If we could measure that heat after the ball stops, we'd discover that it's the same exact amount of energy as went into raising the ball to your chin in the first place.

2 The Second *Law* of Thermodynamics

Remember, the First Law says energy never just goes away. If it seems like there's some energy missing, it has probably become heat. The Second Law explains why the ball stops swinging, and why a ball rolled along a smooth surface eventually comes to a stop even if there's nothing in its way—because all that swinging and rolling energy turns into heat after a while! When scientists came up with this idea, it was a big deal.

Whenever we use energy, some of it becomes heat.

Energy Conversion

Get a rubber band, about 10 cm (4 inches) long. Touch the rubber band to your lips. Your lips are very sensitive to heat, so you can feel how warm the rubber band is.

Now stretch the rubber band over and over again in your fingers about 50 times. That uses a lot of energy, doesn't it? Touch the rubber band to your lips again. It's warmer. The rubber band hasn't gone anywhere. It hasn't changed size much. So all the energy you put into the rubber band doesn't do anything except make it warmer. That energy you put in it has been converted to. . . heat. Yes, heat.

But what can you do with a hot rubber band? That's the problem with the **Second Law of Thermodynamics**. It says that once energy becomes heat, it's very hard to get it back again into a type of energy we can use.

Entropy

Think about the Second Law a little. It means that no matter how hard you try, you can never get all the energy out of an engine that you put in, since some of that energy is going to turn into heat. In 1850 a German scientist named Rudolf Clausius came up with the idea that although energy is always conserved, the amount of "**useful energy**" is always getting smaller. Some of that useful energy is lost to heat.

RUDOLF CLAUSIUS

1822-1888. Came up with the idea of entropy.

Clausius invented a brand new idea which he called "**entropy**" [ENN-truh-pea]. He made up that word from Greek. It means roughly "what everything turns into." Entropy is often described as the amount of "**disorder**" of things. Everything in the universe is tending towards becoming more and more disordered. There's more and more entropy all the time, with less and less useful energy.

The Second Law and entropy have some pretty far-reaching ideas wrapped up in them. If the Second Law is really a law of nature, then the Earth and the Moon and everything in the universe will eventually just slow down and stop. Everything everywhere will just run out of useful energy because that energy will become heat. And since the universe is so huge, the energy would be so spread out that the whole universe will just be a cold, dead place. This may or may not be true. Scientists are trying to understand this all the time. Whatever is going to happen, it's not really something we'll be able to see for several trillion years.

We do know that the Second Law of Thermodynamics is true here on Earth. Here's an example of how we know it's true. Imagine a pond in the summer time. The water has some energy because it's warmer than, say, your freezer or the North Pole. The molecules of the water have energy—they're flowing around, bouncing off each other. Now, have you ever seen a pond like this suddenly freeze on a hot summer day? Well, no! Why not? Because heat always flows from hot to cold. Never, never does heat flow from cold to hot—never. You know this in a way, but you might not have ever thought about it. The energy of the pond water's molecules never flows completely out of the water into the air, unless the air is colder and has less energy than the water.

Here's another classic example of the Second Law: Imagine blowing up a balloon. Now, let go of the end. What happens? The air flies out. The balloon never just suddenly gets bigger, with air flowing in by itself. Never. The air tends to spread out. As far as scientists can tell, everything tends to spread out. Think about spilled milk. It goes all over the place. So a gas in a balloon is thought to have less disorder, or entropy, than a gas that's just free to roam around the universe.

You've probably heard of Murphy's Law: "Everything that can go wrong will go wrong." Like, if it can, a glass of soda will get knocked off the table and break. Murphy's Law is a joke, in a way. But maybe it's our way of dealing with the Second Law of Thermodynamics: Everything becomes more and more disordered—more messed up.

Things in nature may, in a sense, become more disordered. But there's one important catch to the Second Law: It was invented for closed systems—systems of valves and gears and so on that make up steam engines and electric motors. People seem to be able to overcome disorder from time to time—like when we build a good-looking new building or clean up our rooms. We have to put more energy into those jobs, that's all. So don't say you couldn't finish your homework because of the Second Law of Thermodynamics. That would be a lame excuse.

The Third *Law* of Thermodynamics

We're supposed to be talking about thermodynamics—the flow of heat. So lets look at what happens when heat flows away from something. Let's go back to our water balloon in the freezer. Remember how it shrank as it got colder? The water vapor turned from gas into liquid. The air **contracted**. Since everything had less heat, the molecules weren't vibrating so much.

What if we took even more heat out of the water balloon? All the water would freeze at 0°C (32°F). The gasses in the air would contract, getting smaller and smaller. Eventually, the gasses would start to turn into liquids. For instance, if we cool oxygen down to 183°C below freezing (that's -297°F), it turns into a liquid. That's right. The oxygen we breathe can be made so cold that it stops floating in the air and turns to a liquid. It's clear blue, and it pours about like liquid water. Many rockets, including the U.S. space shuttle, use liquid oxygen to mix with their fuel.

Meanwhile, we're still taking heat out of the water balloon. The atoms of liquid oxygen and the other liquefied gasses are vibrating slower and slower. Is there a temperature so cold that there's no vibration at all? All atoms, all molecules—totally still? This is a mighty interesting question. The answer is, as far as we know, that absolutely stopping the motion of the atoms and molecules would be as cold as we could ever get in this universe. This temperature, this absolutely coldest temperature, is called "**absolute zero**" because there's zero vibration. It's -273°C, or -460°F.

What would matter be like at absolute zero? Really cold—extremely cold! All gases and liquids would be solid. All the atoms and molecules would be packed close together because as things get colder, they get smaller. And none of those atoms and molecules would be vibrating. Their energy would be zero. Sounds like a pretty weird and even scary place. That brings us to the Third Law of Thermodynamics:

At absolute zero, all motion stops, but we can never quite get there.

TEAM ZERO

Scientists try to take molecules down to absolute zero because slower molecules are easier to study. But we can never quite get to absolute zero. See, whatever stuff we're trying to get that cold must be connected somehow to the stuff in the lab. For instance, suppose we had something in a special refrigerator that was set for absolute zero. Well, the refrigerator must be sitting on the lab floor. The floor and the whole lab are not at absolute zero; they're really warm in comparison. So whatever is almost at absolute zero is always going to be ever so slightly warmed by something it's connected to. Even if the whole lab was almost at absolute zero, there would be heat coming from somewhere in the universe. We can get very, very close, but never quite there. We've gotten to within a thousandth of a degree!

A British scientist named William Kelvin gave this idea of absolute zero a lot of thought. He realized that if you started a temperature scale at absolute zero, you'd never have to mess around with negative numbers, like "forty below zero" (-40°F). The temperature would always be positive. For this, he got another temperature scale named after him. On that scale 273 **Kelvins** is the freezing point of water, the same as 0°C, or 32°F. To be real cool, scientists (like you) just say, "273 Kelvins." We don't say "273 degrees Kelvin." So what's the coldest temperature scientists have ever reached? How about around .001 Kelvins!

WILLIAM KELVIN

1824-1907. Based temperature scale on absolute zero.

Moving Heat Around

There's another very important part of studying heat called "**heat transfer.**" It means how heat moves from one place to another, like from a stove to a meat loaf, or from the air to your body on a warm day, or from the Sun to the Earth.

Conduction

The first way heat moves is very simple: One molecule bumps into another. That's right. They transfer heat energy by smacking into one another. They act like a good night on the bumper cars. Suppose you put your finger in something warm, like a cup full of hot cocoa. You feel heat instantly, because the vibrations of the hot molecules in the cocoa are bumping right into the molecules of your finger. The bumping molecules transfer energy. Your finger ends up a little warmer, and the cocoa ends up a little colder, with less heat energy.

This sort of heat transfer is like when you push a cue ball into another pool ball. Energy is transferred from the cue ball to the other ball. The cue ball slows way down and almost completely stops. The other ball, meanwhile, goes rolling off with just a little less energy than the cue ball had. Well, that's what happens with molecules. Molecules transfer their heat energy to other molecules when they bump. This way of transferring heat is called "conduction" [kon-DUK-shun].

Thermometers work because of conduction. The air whose temperature you're measuring surrounds the thermometer. Its molecules hit the molecules of the glass tube, if that's the kind of thermometer you have. The glass molecules hit other glass molecules until the innermost glass molecules hit the mercury atoms. Then the mercury starts to expand, showing you the temperature rise. It seems simple, and it is!

Conduction

Here's What You Need:

- *3 or more knives or spatulas made of different materials. See if you can find a metal knife, a wooden spatula, and a plastic picnic knife.*
- *Big mug*
- *Hot water*
- *Pats of butter. Sure, margarine will work just fine.*
- *Sugar cubes*

Here's What You Do:

1. *Stick a dab of butter on the end of each knife or spatula, and then place a sugar cube on the butter. Looks a little weird, but keep going.*
2. *Fill the mug with hot water.*
3. *Put the handle of each knife or spatula in the hot water, with the butter and sugar leaning out from the mug.*
4. *Watch to see which sugar cube comes loose first.*

Here's What You See:

The metal knife's sugar cube drops off first because the butter that was holding it on gets soft. It takes a long, long time for the butter to soften on the plastic and wood utensils.

What does this show? Well, by putting all the knives in the same hot water, you make the submerged ends all the same temperature. Then, whichever substance warms up the butter fastest wins the drop-the-sugar-cube race. And how does the butter get warmed up? By **conduction**, see? First the heat energy is transferred up the knives. Then it's transferred from the knives to the butter.

Not all materials conduct heat the same way, even though they're all made of molecules and atoms. Metal is the best heat conductor in this bunch. The electrons in metals are easy to push from one atom to another. It's like a field full of beehives, and the bees can just roam from one hive to another (at least in your mind's eye). The hives stay put, but the very small bees can move from one to another easily. So it is with electrons in metals.

Insulation

Along with conduction goes another idea, "insulation" [in-suh-LAY-shun]. Insulation comes from the Latin word for "island." When you're insulated, you are cut off from heat or cold, just as an island is cut off from water. Well, maybe not totally cut off, but that's the idea anyway. A blanket on your bed keeps you from getting cold. It insulates you. Insulation makes the transfer of heat energy between the particles of matter happen slowly. It's as if our pool balls from before are running into balls made of silly putty or soft clay. Although the second ball gets hit, it doesn't shoot off with the same quickness that a hard pool ball does. It's the same with materials that insulate. Heat doesn't move very quickly through them. **A good conductor is a bad insulator**, and the other way around.

Air is a very good insulator. Since air is a gas, its molecules are much, much farther apart than in a solid like metal. With the molecules far apart and having to go a long way to smack into one another, heat energy doesn't move through air molecules very fast.

You may own a down feather vest or jacket. These coats trap air between you and the cold weather. The down feathers hold the air still, and the coat itself holds the still air next to you. See, that's how birds keep warm; they wear very toasty feathers all the time. Here's another interesting fact: caribou and reindeer have hollow hairs in their fur coats. Each fiber of their fur traps still air and keeps them warm. That way they can hang out near the North Pole.

There's something that insulates even better than air—nothing. In other words, a **vacuum**. No matter at all! Heat can't be moved around by bumping molecules and atoms if there are no molecules or atoms to bump. That's how a thermos bottle works. It's like one bottle inside another, with a vacuum in between.

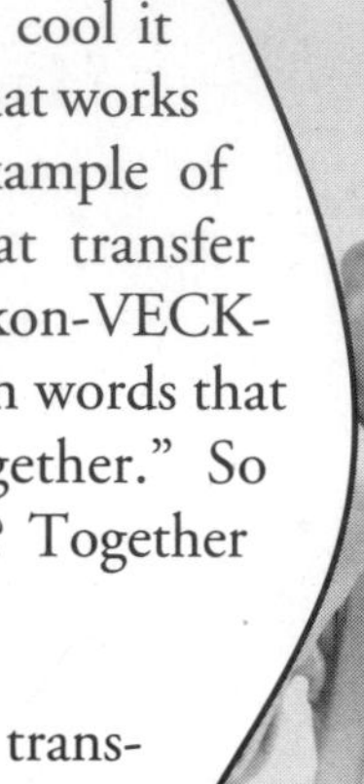

Full of Hot Air—Convection

Have you ever eaten soup that's just too hot to have on your tongue? Ouch! Did you cool it off by blowing on it? That works a little. That's an example of another type of heat transfer called "convection" [kon-VECK-shun]. It's from Latin words that mean "to carry together." So what's being carried? Together with what?

Well, convection is heat transfer from or to something that can flow, like water or soup. And it doesn't have to be a liquid. Air flows too. So does helium, and hydrogen, and any gas. Things that flow are called "**fluids**" [FLOO-idz]. And all fluids can transfer heat by convection.

Now, back to our example of the hot spoonful of soup. If you blow on a spoonful of soup to cool it, the somewhat cool air molecules of your breath pick up heat from the soup molecules by conduction. Then the air molecules flow away, carrying the heat along with them. Those air molecules spread the heat out into the air around you by convection. The movement of the air when part of it is heated sets up what we call a "**convection current.**"

COOL!

Convection Current

Here's What You Need:

- *Glass dish of the sort you can heat on the stove. One type of this heat-resistant glass is called Pyrex.*
- *Water*
- *Stove*
- *Food coloring*
- *Pot holder*

Here's What You Do:

1. *Fill the glass dish half full with water.*
2. *Place it on the burner of a stove so that the container is only half on the burner. Support the other half of the container with a pot holder.*

3. *Turn the stove on medium or "medium-low," if your stove has that setting.*
4. *After five minutes, put a few drops of food coloring in the water near the edge of the fire-resistant glass.*

Here's What You See:

Follow the drops of food coloring with your eyes. You see, almost right away, that the water over the burner is rising and being replaced by cool water from the side off the burner. That's a convection current in action. It's pretty cool.

How does the convection current start? As the water warms, its molecules spread out. When there's more space between them, the molecules become less dense. That means the water over the burner weighs less for the amount of space it's taking up. There are actually fewer water molecules in the half of the dish over the burner than there are in the half away from the burner. The cold water, the water not on the burner, gets pulled down by gravity because it weighs more for the amount of space it takes up. It forces the warm water up. Then some of the cold water on the bottom of the dish near the burner gets warmer, and some of the hot water at the top loses heat to the air. This cool water sinks and pushes the new warm water up. This constant place-switching is a **natural convection current**.

You may have heard the expression "hot air rises." That's right, it does. So does hot water. But the reason hot things rise is not just that they're less dense. It's also that the surrounding fluid is more dense. Gravity pulls the cooler fluid down, forcing the warm fluid up—naturally.

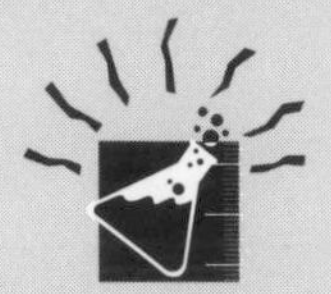

Hot Air Balloon

Here's What You Need:

- *Blow dryer*
- *Extension cord so you can take the blow dryer outside on a nice day.*
- *Plastic trash bag, the thinner the better. The crinkly, very thin bags used to line trash cans work well.*
- *Tape. Regular transparent adhesive tape will do.*

Here's What You Do:

1. *On a nice dry day, plug the blow dryer into an extension cord and take everything outside.*
2. *Pinch the open end of the bag so that it will fit loosely around the outlet or nozzle of hair blow dryer. Tape it, using as little tape as you can (you don't want the bag to get too heavy).*
3. *Loosely fit the bag over the end of the blow dryer. Be sure not to squeeze the bag too tightly around the blow dryer. The air has to have a way to get out, or you'll overheat the blow dryer!*

WARNING!

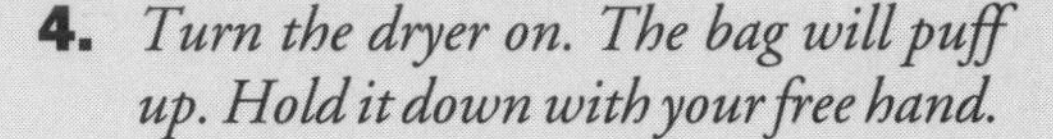

4. *Turn the dryer on. The bag will puff up. Hold it down with your free hand.*
5. *When the bag is warm to the touch on top, let it go. The bag will go up. Turn off the blow dryer. With the right bag and blow dryer, these hot air bags will go 20 m (60 feet) into the air—way over a house!*

Here's What You See:

Our hot air balloon, which was once an ordinary trash bag, gets filled with air heated by the blow dryer. Then the less dense hot air inside the balloon is forced up by the cold air surrounding the balloon. And away it goes!

"**Forced convection**" means that we force, or push, air or another fluid in order to transfer heat. When you blow on a spoonful of soup to take heat away, that's forced convection. When a hair dryer blows out hot air, that's forced convection.

Radiation

Hold you hand *under* a light bulb, about a hand's width away. Does your hand feel hot? Now, how could that be? You aren't touching the light bulb, so there's no conduction. The convection current makes hot air rise from the light bulb; hot air doesn't sink down from the bulb. So how does heat energy get from the bulb to your hand? This is the type of heat transfer called "radiation" [RAY-dee-AY-shun].

Radiation is heat traveling as **infrared** [INN-frah-RED] energy. (More on how it got that name in the next chapter—just wait!) Infrared energy moves through the universe at the speed of light. It can even transfer heat in a vacuum, which conduction and convection can't. Just as some types of matter are better heat conductors than others, some things catch heat radiation better than others.

Absorbing Heat Radiation

Here's What You Need:

- *Two identical boxes made of corrugated cardboard*
- *Black paint or black construction paper*
- *Aluminum foil*
- *Your hand*

Here's What You Do:

1. *Paint the inside of one box black and let it dry, or line it with black paper.*
2. *Line the second box with aluminum foil. Put it next to the first box.*
3. *Put your hand inside the black box. How does it feel? Then put your hand inside the foil-lined box.*

Here's What You Feel:

Your hand feels much warmer in the middle of the aluminum. The shiny surfaces in the aluminum foil box reflect heat back to your hand. The black surfaces absorb heat from your hand, not allowing it to reflect back to your skin. Isn't that wild? There's no source of heat in there except your hand. There's no heater or ice or anything. That's just the heat of your hand. You're feeling heat transferred by infrared radiation.

Observing cars is an ideal way to study infrared radiation. The insides of dark colored cars, particularly black ones, get much hotter on a summer day than light colored ones. The black paint **absorbs** *more heat than light colored paint.*

Sending Out Heat Radiation

Sit in a chair in a room with an uncovered window (no curtains or shades) at night. Get relaxed. Now hold your hand so that your palm faces out toward the window. Concentrate and feel how cool your palm is.

If it's cool out at night, look at car windshields in the morning. The windshields that were covered by a carport roof or a tree often don't have frost or fog on them. The windshields that had nothing over them during the night are often frosty or foggy. Now why is that? It's because a car windshield, or anything else that's exposed to the night sky, is really being exposed to **outer space!** *And outer space goes on for just about forever. There is nothing to stop the heat*—**nothing!**

Rotate your palm so that it faces in toward the room. Does it feel warmer? You can really notice the difference. Your hand is sensitive enough to feel when you're radiating heat right out the window into space!

Your hand is just like a flashlight, except that it radiates heat. Okay, imagine instead of using the palm of your hand, you were using an ordinary flashlight. If you shine the flashlight toward a wall in the room, some of the light will reflect back to your eyes, and you'll be able to see it. If, on the other hand, you were shining the light out the window, where would it go? Off into space. Off the Earth—gone forever. The same thing happens to the heat from the palm of your hand. It is either reflected back to you by the walls of the room, or it goes right out the window.

Here Comes the Sun

Heat can really zoom along in outer space. As the Sun gives off energy, there's nothing to stop it as it zips 150 million kilometers (93 million miles) to the Earth. That's a long, long way, but we still feel the heat from the Sun every day. From 150 million kilometers away!

This business of **solar radiation** greatly affects how we design our satellites. In space, things that face away from the Sun get extremely cold. And any part of a satellite facing the Sun gets very hot. There is nothing to keep the satellite cool on that side. So scientists often build satellites with sun shades which look just like fancy silver beach umbrellas. We also often set satellites spinning very slowly so that the heat from the Sun is spread out a little.

The Sun warms the air with its radiation, and the air spreads that heat around by convection. In fact, the Sun's radiation is the source of most of the energy we have on Earth. The Sun's heat makes the wind blow (more about that in Chapter 8). It helps plants to grow, and animals (including us humans) get energy from eating those plants.

LIGHT

chapter five

Look all around you. What do you see? Now that you've read Chapter 2, you might think, "Matter!" But actually you see the *light* coming from different forms of matter. Like heat radiation, light is a kind of energy traveling through space. It's the sort of energy that our eyes can see.

The S p e e d of Light

Light travels fast. Very, *very* fast. How fast? Well, it goes "the speed of light," of course. That's 300,000 kilometers per second, or 186,000 miles per second. *Per second.* Not per minute, not per hour, but per second! Light, if it is properly directed, can go all the way around the Earth more than seven times in just one second. That is amazingly fast. It's cool to think about how fast light travels:

Where you are:	Source of light:	Time to get there:
Home plate	Outfield fence 125 m (410 ft)	1/2 of a millionth of a second
Soccer goal	Other goal 100 m (328 ft)	1/3 of a millionth of a second
Earth	Moon 376,280 km (233,810 miles)	1 1/4 seconds
Earth	Sun 150,000,000 km (93,000,000 miles)	8 1/3 minutes
Earth	Proxima Centauri, the nearest star 41,000,000,000,000 km	4 1/3 years

There are a couple other important facts about the speed of light which Albert Einstein figured out. (Einstein did a lot of thinking about the speed of light. Remember his formula $E=mc^2$? The "c" is the speed of light.) The first important fact is that the speed of light through empty space never changes. It's constant. That's why we use the letter "c." Near as we can tell, nothing in this universe can go faster than the speed of light. So remember that universal speed limit:

SPEED LIMIT 186,000 MILES PER SECOND

300,000 km/second 186,000 miles/second

When you're playing ball or riding in a car or even flying in a fast airplane, light seems to go from one place to another instantly. Our eyes can't see how fast light travels. But it still takes some time to get from one place to another. In fact, you can get an idea of how slow light travels while watching television.

experiment

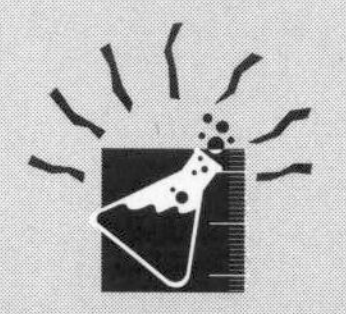

The Slowness of Light

Find a news show on television. Watch until the announcer in your country is talking to a reporter on another continent. The announcer might tell you that he or she is doing a "satellite" interview. Then listen for pauses between the questions and answers of that interview.

The person in the far away place waits and listens for the announcer here to finish talking. If you watch carefully, you can see that the person far away seems to take a few instants to start answering the person in your country. That's because the television signal needs time to travel. It goes the speed of light, so that time is a very small amount of time. Nevertheless, you have to wait for the signal to get from here up to a satellite and back down to the other side of the world. Then the far away person's signal has to go back up to the satellite and back down to Earth for you to see it. Once you notice newspeople trying to carry on conversation like this, you can tell right away whether or not the broadcast is live.

Satellites used for these broadcasts are about 36,000 km (22,000 miles) above the Earth. At the speed of light, it takes the television signal about a quarter of a second to travel from place to place. Try watching TV with stopwatch, and see if you can measure the quarter of a second. It takes time for light to make the trip around the world—not very much but some.

What Meets the Eye

Whether it's traveling from a satellite to a TV studio or from this page to your eyes, light travels in straight lines. That means you can see light only when it comes right into your eye.

experiment

Beam of Light

Go into the kitchen and turn off all the lights. Shine a flashlight along the kitchen counter. Can you see the light beam? Not until it hits the wall, probably. Now sprinkle a little flour in front of the light. When you do that, you see the light beam coming out of the flashlight. Actually, what you're seeing is light coming out of the flashlight and bouncing off the flour grains into your eyes. The light was traveling through that space before, but you couldn't see it since it wasn't pointed at your eyes.

Now turn on the lights. See everything clearly now? That's because the light from the kitchen lamps is being bounced all around by all the other things in the kitchen: the walls, the floor, the shiny appliances. Some of that light bounces off the flour grains and counter straight into your eyes, showing you what you should clean up now!

Reflection: Bouncing Light

When light is streaming past us, we don't see it. We only see light when it travels straight into our eyes. And that can create some pretty weird sights. Scientists, like you and me, call this bouncing "reflection." You probably already know this word from when you look in a mirror and see your reflection. Look at a mirror. Yes, right now. Almost all of the light that hits the mirror is reflected back in your direction, so you see yourself clearly and in full color.

Now look at a wall. Can you see yourself? Can you see anything but wall? Yes, you see light being reflected off the wall. After all, the wall isn't making that light by itself. The light must come from lamps or the Sun. It bounces off the wall, just as light bounces off a mirror—but not all of the light this time. In fact, the only light that is reflected off the wall back to your eyes is the part of the light that is the color of the wall. If you're looking at a blue wall, the wall is reflecting the blue light and absorbing the rest.

Reflections can send some weird sights to our eyes, as the next experiments show.

Where's the Rest of Me?

Here's What You Need:

- *Drinking glass with a smooth, clear bottom*
- *Finger*
- *Table*

Here's What You Do:

1. *Carefully place the glass near the edge of the table so that you can look up through it.*
2. *Stick your finger in the glass so that its tip is just below the water's surface.*
3. *Look up at the glass from below the level of the table. How much of your finger can you see?*

4. *Carefully lift the glass of water so that it's directly over your head. How much of your finger can you see now?*

Here's What You See:

When you look through the glass from below at an angle, you see only see the

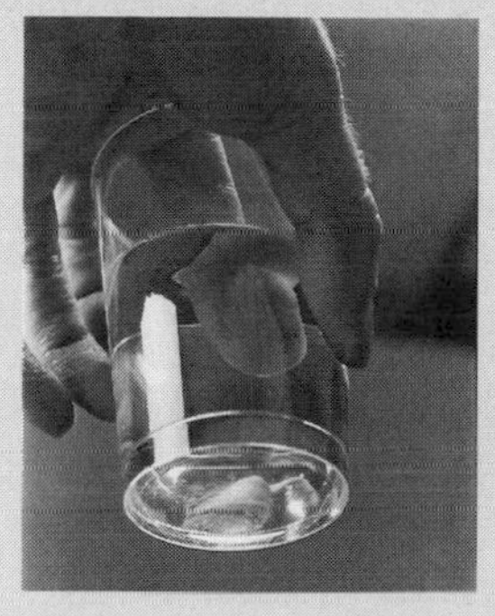

tip of your finger and its mirror image. Where's the rest of your finger? It's still there, but the water doesn't let the light bouncing off it come down at an angle that would reach your eye. You only get to see light that is reflected off the bottom of the surface of the water.

When the glass is over your head, you can see all of the light coming through. And you can see all of your finger. But you can't see the reflection from the bottom of the surface of the water anymore. The light coming through the surface is making it to your eye when your eye's in the right place.

Here's another rule about mirrors and other reflective surfaces: The light bounces off a mirror at an angle that's equal to the angle that the light hit the mirror with. If you're looking straight at a mirror, the light comes straight back into your eyes. If you look at a mirror from an angle, you see the light coming into that mirror at an angle. That causes some crazy sights when the mirror is curved!

Playing with Spoons

Get a metal spoon—the bigger and shinier the better. A good soup spoon is best. Look at the back side—the side that won't hold any soup. Your nose is a little big, but you look like you. Now, turn it over so that you're looking at the scoop side. You're upside down!

To see why this happens, think of lines of light traveling to the spoon and back to your eye. Since the spoon is curved, the light hits it at an angle. That means it bounces off at an angle, too. The angle is sharp enough to make the top and bottom cross on the reflected light's way back to your eye.

Refraction: *Bending* Light

If something lets light go through it pretty well, we say it's "**transparent**" [tranz-PEAR-ent]. That's from the Latin words for "to appear through."

HAHA!

My brother has a device in his house that lets him see through walls! He calls it a "window."

Well, I've got something between my house and my neighbor's house that lets me see her clearly. It's called air!

What these jokes mean is that glass and air are transparent. So is water. But light usually doesn't get through any transparent matter without being changed. It usually changes speed, just a little. When light changes speed, it can change the direction it's going as long as it hits the transparent material at an angle. Changing light's direction—bending light—is called "refraction" [ree-FRAK-shun].

Lend Me a Lens

When we talk about refraction, we're talking about controlling light—controlling pure energy. If you wear glasses, you're doing that all the time. The lenses in your glasses are refracting light before it hits your eyes.

WHOA!

experiment

Refraction

If you can find a magnifying glass, hold it over the light bulb in a lamp. Look at the ceiling. If you hold the glass at the right distance from the bulb, you can read the printing on the end of the bulb right on the ceiling. The light going through the label is going through the lens and being spread out so that it takes a lot of room up on the ceiling.

Think about this, too. The closer the ceiling is to the bulb, the smaller and brighter the image is. If you could move the ceiling farther away, say by (carefully) putting the lamp on the floor, the image would get bigger; but it would also get dimmer. The light making the image is more spread out.

If you wear glasses, you know what we're talking about. With glasses, we can make light come into our eyes just the way we want it too. Glasses have lenses that are made of glass or plastic. Light goes through these transparent materials about two-thirds as fast as it goes through air. So, we can redirect or "bend" light quite a bit.

Bending light is how telescopes and microscopes work. They gather light and redirect it so that the light reflecting off a small portion of whatever we want to see takes up our whole field of vision. In the case of a microscope, the light is bent so that some tiny feature, like an ant's face, takes up all our whole field of view. A telescope takes a tiny portion of the sky and makes it take up our whole field of view.

MARCHING BAND OF SCIENCE!

The Light Wave Marching Band

You've seen marching bands; you may even have marched in them. Think of light as lines of marchers in a marching band. Imagine those marchers walking along a nice paved street; that's like light zipping through air.

Then the marching band comes to the end of the street and steps onto a muddy field. They have to slow down. It's impossible to walk so quickly when the ground is slippery. (Remember these people are marching, not sliding or taking head-first dives into the mud, at least until the experiment is over.) Okay, so the waves of marchers slow down as they step in the mud. That's fine if every marcher in the first line hits the mud at once. They all go a little slower, but the band ends up marching out of the muddy field in the same formation.

But imagine the marchers coming onto the muddy field at an angle. The marchers who hit the mud first would slow down sooner than their buddies, who walk a bit farther before getting their feet wet. What happens? The line of marchers ends up turning. Isn't that wild? As one edge of the marching "wave" slows down, the other edge keeps going and the whole line turns. In fact, that's how marching bands turn. When, they come to a corner and need to turn, the people on the inside of the turn slow down on purpose, and the outside people take long strides around the outside edge of the turn.

Keep this idea going in your head for a moment. Each "wave" of marchers that hits the muddy field at an angle slows down on the edge closest to the mud before the people on the side away from the mud. And the whole line turns. Wave after wave turns. The same thing happens with light. As a light wave hits the surface of material that cause refraction at an angle, it slows down on one side, and the whole wave changes direction.

Notice too, that it works as the wave comes out of the material. Because the wave is coming to the surface at an angle, part of it gets out first. That part then starts moving faster first, and the wave changes direction. Going back to our marching band, if the band got back onto a paved street or parking lot, and they did it at an angle to the pavement, they would change direction. After you figure it out, you think, "Yeah, I knew that; how could it be any other way?" It's pretty cool.

Another Lens

You can make a bag of water into a magnifying lens. The idea is to get water, which slows light down, into a curved shape.

experiment

Water Lens

Get a clear plastic sandwich bag that closes with a "zipper." Fill it pretty full with water and close it tightly. The bulging sides of the bag make this a lens. Obviously, you should be careful with it. It's a bag of water; you can make a big mess in a second.

Anyway, hold this water lens up to some object you have around, like your other hand. You'll see it's magnified. You have a water bag magnifier. What a party! The more your water bag magnifier is curved, the more it will magnify—try it.

The plastic bag itself bends light, but it's so thin that it doesn't affect the light's path much. But light goes through water about three-quarters as fast as it goes through air. So water bends light like glass, although not quite as strongly. (Water doesn't slow light down as much as glass does. To think back to the Light Wave Marching Band, it's as though glass is a very goopy field, and water is a field that's not quite as goopy. So our wave marchers would not be slowed down as much on a water field as they would on a glass field, and they would, therefore, not bend as much. Whew.)

Index of Refraction

If you were designing eyeglasses or a telescope or a camera lens, you'd want to know just how much plastic or glass bends light. Well, scientists have worked on this problem for a long time. They've come up with a way to tell how much light would bend in transparent materials. It's called the "Index of Refraction." Index comes from the Latin word for "first finger." So an index is something that "points out with your first finger." The way you might look up something in a book. Not bad.

The index of refraction of water or glass is the speed of light in a vacuum divided by the speed of light in the water or glass. In a vacuum there's nothing to slow the light down. No air, glass, plastic, or water. Nothing.

Substance	Index of Refraction
Vacuum	1
Air	1 3/10,000
Water	1 1/3
Plastic bag	1 1/2
Glass	1 1/2
Diamond	2 1/2

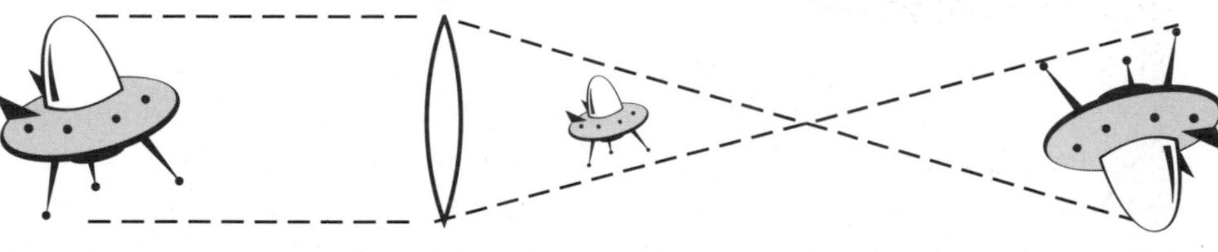

Find the Focal Point

Get out your magnifying glass or your plastic bag full of water again. Move some object, like this book, close to and then away from your water lens. You can see the object magnified. As you move the object farther away, you see a big blur; then it's upside down! Isn't that wild? The light is bent by the lens. So at some point, the light will bend in such a way that it crosses over itself. That point is the "focal" [FO-kull] point. Well, when the object you're viewing is on one side of the focal point, it's right side up and magnified. When it's on the far side of the focal point, it's upside down and magnified a different amount. All this because the water is refracting the light, so that when the light hits the water at an angle, and the water slows the light down a little.

If you shine a light through a lens, the focal point is the place where all the light comes together, or "**converges**" [kohn-VERJ-ezz]. Of course, this is for a well made, carefully shaped lens, like you find in a good pair of eyeglasses or a nice desk magnifying glass. For our bag of water lens, the point isn't a point so much as a region. It's not bad, though.

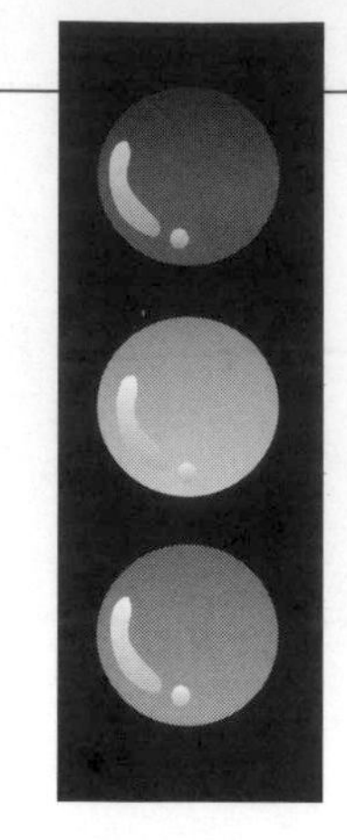

Red Light, Green Light: Colors

So now you know that light comes in beams that can be bounced around and bent in different ways. But here we get really wild. Those light beams can also be *split* into parts. And the parts of white light are colors. We call the colors the "spectrum" [SPEK-trum] of light. What we see as white light is actually a mixture of all the colors of the rainbow—the full spectrum.

experiment WILD!

Make a Light Spectrum

Here's What You Need:

- *Small mirror*
- *White wall where the Sun isn't shining, or a white pad of paper set up where the Sun isn't shining*
- *Pan, like a brownie pan*
- *Water*
- *Sunshine or other bright white light*

Here's What You Do:

1. *Fill the pan with water and put it in the sunlight next to the white surface.*
2. *Hold the mirror at one end of the pan so it gets hit by the sunlight.*
3. *Shift the mirror's angle until you see a spectrum of light on the white wall.*

Here's What You See:

When the sunlight falls on the water and mirror at just the right angles, it breaks into the spectrum of colors. Now remember that since you're breaking light up into the different colors, the beams of colors will each be only partly as bright as the white light. So in order to see them clearly, make sure your "viewing screen" (the white surface) isn't directly lit by any light source. Keep it in the shade.

You can also make a spectrum with cut glass. Check with your parents, and find a glass object that has somewhat sharp edges. Often the stem of a fancy glasses will be cut sharply, or the pieces that dangle from chandeliers. Hold the glass in sunlight. Hold the glass object so that some of the light that passes through the glass lands on a white surface. Turn the glass until you see all the colors of the rainbow.

A **prism** *is a triangular piece of glass. They're not complicated-looking. They're nice evenly shaped objects that will break light into colors.*

Colors of the Rainbow

So the spectrum is all the colors of the rainbow. Okay, then what is a rainbow? Instead of glass or plastic prisms, rainbows come from. . . what? Well, from rain. Raindrops act like very small prisms. When rain is falling from high in the sky, the droplets are round. The light goes in one side of the raindrop, bounces off the back side of the raindrop, and comes back out the front broken up into colors—just like a prism.

To see a rainbow, your eyes, the rain, and the Sun have to be lined up just right. The Sun has to be behind you, shining into rain that's falling off in the distance. The Sun has to be low enough in sky for the light to hit the raindrops and come back to you at an angle of almost exactly 42 degrees, because of the way water bends light. The light comes from behind us into the raindrops and back toward our eye.

Now, you may see all sorts of things that are just like rainbows. Near a waterfall or lawn sprinkler you can often see a small rainbow caused by water droplets sprayed into the air from the splashing of the falls or the sprinkler. Sometimes if you are in an airplane looking in just the right direction, you'll see a small completely round rainbow, caused by the water droplets or tiny ice crystals in clouds. They all work the same way: the water droplets break the light up into the spectrum of colors and reflect them toward your eyes.

Every person who sees a rainbow is seeing his or her own rainbow. It's slightly different—slightly off from anyone else who's also seeing it. In a way, rainbows occur every time it's raining and the Sun is shining, but you just have to be in the right place to observe them.

Catch a Light Wave, Dude!

Seeing the different colors of the spectrum is one thing, but where do they come from? What makes green light, say, different from red light? In many ways, light acts like it's made of waves zipping along. You've felt waves whenever you've been on a boat. There are two things going on with waves: They're moving forward, and they're going up and down. All the time, waves are moving in two directions at once.

So think about this: light waves are all going forward at the same speed (the good ol' speed of light), but they're going up and down at different rates. Some are going up and down fast, some slow. The different colors come from the light waves going up and down at different rates. We say that each color is a light wave of a different "**frequency**" [FREE-kwenn-see].

Here's a good way to think about frequency: Imagine yourself standing on a ship anchored at sea. Like any scientist, you're interested in waves, so you watch them. As they go by, you feel the boat go to the low point of the wave. Then you feel the top of the wave. Okay, now you could, and I hope some time you will, count the number of tops that go by every second. The time between tops is called the waves' "frequency." (You can also count bottoms; the frequency is exactly the same.)

How do light waves have the same speed but different frequency? Let's go back to the Light Wave Marching Band. Let's say the marchers are just walking now, not marching in step, but they're all moving at the same speed. Some take shorter steps than others. And the ones that take shorter steps have to take more steps in one second than their long-legged buddies so they can keep up. They're not marching anymore, but they all get to the same place at the same time.

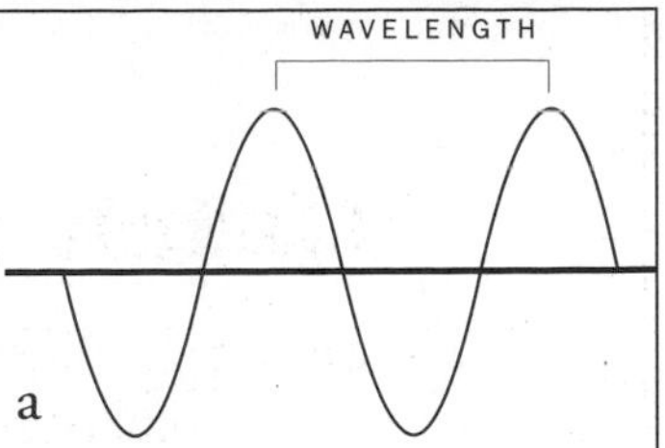

WAVELENGTH

Frequency and Wavelength

You can measure light waves in two ways: frequency and wavelength. The **frequency** is how many waves come through in a second. The **wavelength** is how long each wave is. That makes sense: wavelength = length of the wave. (Not that tough.)

Here's the important rule: **the longer the wavelength of light, the lower its frequency.** And the shorter the wavelength, the higher its frequency. That makes sense, too: If the waves are shorter, more of them can go by in one second.

For visible light, when the frequencies get lower, we see the light as red. And when the frequencies get higher, we see the light as violet. That puts yellows and greens somewhere in the middle. So when you see red light, it's traveling at the same speed right along with blue light, but red's waves aren't going up and down as fast.

Light and Suds

You might wonder how we humans figured this out. Well, it took a long time—a few centuries. Scientists conducted all sorts of experiments, like Isaac Newton's experiment with prisms. The next experiment is another you can do with soap bubbles. No kidding—soap bubbles.

Calling interference

Here's What You Need:

- *250 mL (1 cup) dish washing soap*
- *750 mL (3 cups) water. Soap and water are all you really need, but you can try adding these optional ingredients as well:*
- *Glycerin [GLISS-er-inn]. Sometimes it's called glycerol [GLISS-er-all]. You can get it at the drug store in the skin care section. It's a natural oil.*
- *Corn Syrup*

Here's What You Do:

1. *Mix the dish washing liquid and the water into a smooth liquid.*
2. *If you live where the air is dry, add the glycerin. You need just a little: about 15 mL for every liter, or 1 tablespoon for every quart.*
3. *If you live where the air is humid, you might want to add a little corn syrup: about 15 mL for every liter, or 1 tablespoon for every quart.*
4. *Blow some bubbles outside on a day when it's sunny; or make some bubbles in the kitchen with bright lights on. Look at the bubbles.*

Here's What You See:

In the bubbles you can see colors. Watch closely; they're changing all the time. If you look closely, you can see the liquid in a bubble flowing around its surface.

So here's the question: Where do these colors come from? The sunlight shining through the bubbles is white. But you're seeing the same colors as come out of a prism.

The colors in the bubbles come from the waves of light running into each other. You, as a scientist, can call this "**interference**" [inn-ter-FEAR-inss]. It's the same word you hear the sports announcer say when there's "pass interference" in football. One player runs into another player when they're both trying to catch a pass. Well, light goes into the bubble's outer surface, bounces off the inner surface, and comes back out. The light coming into the bubble runs right into the light coming out. The light waves pass right through each other. While they're at it, they get in

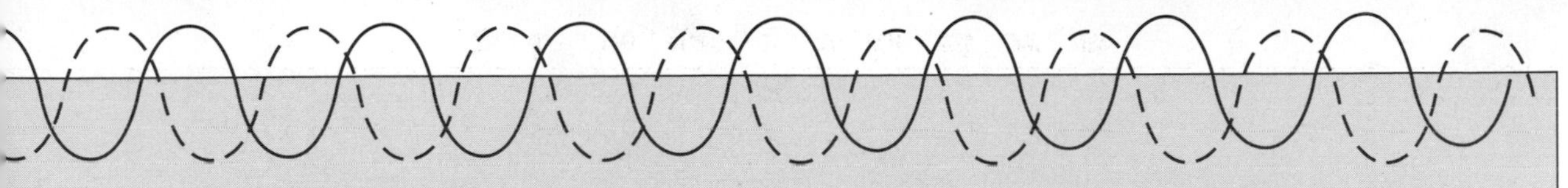

each other's way; they interfere with each other. So what happens, we still see *some* light, right?

Well, the bottoms of one wave pass through the tops of another and cancel them out—make them go away. No bottoms or tops left over means no light waves of that type. Meanwhile, other waves add up; one wave's top adds to another wave's top. That leaves particular frequencies of light coming back, which are the only ones we're going to see. The frequencies we see are the colors on the bubble! Right there!

As the soapy water in the very thin bubble flows and shifts around, the bubble's thickness is changing too—just a little. Now, when we say little, we mean little indeed. The bubble's skin is changing about 100 billionths of a meter (about 40 billionths of an inch)! That's about how long the wavelengths of light are. It's pretty cool to realize that when you look at the soap bubble and see colors, your eyes are measuring billionths of meters or inches.

Roy G. Biv

There's a famous guy in science named Roy G. Biv. He didn't discover anything cool. He didn't explain any mysteries. In fact, he's not a real person. Roy G. Biv is just a trick for remembering the colors of light broken up by prisms, raindrops, and so on. The colors are, in order:

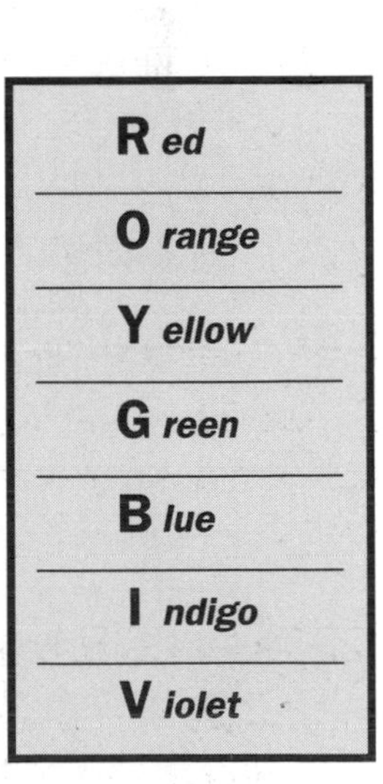

Now there's more to light than ROY G. BIV. Those colors are only the parts of light that we can see. There are types of radiation that have frequencies lower than red, and frequencies higher than violet. In fact, you already know about one of those types of radiation from Chapter 4.

But before we go on, you might wonder: what's with this "indigo" deal? Indigo [INN-dih-go] is a very dark blue, almost purple. It really is there in the spectrum, right between blue and violet. The name of the color comes from the indigo plant that is used to make dark blue dye for clothes. But we don't talk about indigo much. It's pretty, but it's not in the usual Top Ten of colors. It's more like in the Top Twenty-five, down near peach and vermilion. So how come it gets a place in the spectrum?

Well, it's very important to keep green in the middle of the spectrum. Of all the wavelengths of light that we can see, green is the wavelength almost exactly in the middle. So we make sure that ROY and BIV have the same number of letters, with G in the middle. Gee whiz!

Infrared Light

Infrared [INN-frah-red] light is the light just below red. In Latin, "infra" means "below." The wavelength of red light is around 700 nanometers (nm). So the wavelength of infrared light is just above red light, around 720 nm.

Now where have you read about infrared energy before? That's right—it's heat radiation! When heat radiates, it acts exactly like light, except our eyes can't quite see it. Some other animals, particularly owls, can see infrared light. That's how they hunt at night. Mice and other animals give off heat because of their body temperatures. To something that can see infrared light, they actually glow. Scientists have developed goggles that people can wear to see infrared light; they're called "night vision" goggles.

Another way we've learned to use infrared light is in the remote controls for televisions and VCRs. The infrared used in remote controls is very weak, and it's at a very carefully chosen frequency that the TV or VCR is set up to receive.

Infrared Light

Here's What You Need:

- *TV with remote control*
- *Black construction paper or a dark sock*
- *White paper*
- *Mirror*

Here's What You Do:

1. *Point the remote control at the TV and switch it on and off to make sure it works.*
2. *Point the remote control away from the TV, and switch it on and off again. Does anything happen?*
3. *Hold the mirror away from the TV. Point the remote control at the image of the TV in the mirror. Does clicking the remote switch the TV on and off?*
4. *Instead of a mirror, use the white paper. Reflect the infrared waves off the paper toward the television. Can you switch the TV on and off?*
5. *Try the same trick with a piece of dull black paper or cloth. As long as the black material isn't too shiny, the remote won't work.*

Here's What You See:

Pointing the remote control at something black doesn't do anything. The black absorbs the heat from the infrared sender in the remote. But if you point the remote control at the white paper or mirror and have it lined up right, you can control the TV with the remote pointing completely away from it. When the infrared hits the white paper or mirror, some of the radiation is absorbed and lost to heat, but enough bounces off to get the job done.

If you can find an old remote control and your parents say it's okay, take the cover off the remote, and look at the sending end—the front. You'll see a very small device that looks like a tiny light bulb. It is. But this bulb only glows with infrared light. When the remote sends signals to the TV, the infrared light bulb is glowing, but we don't see it.

White paper is a better reflector of visible light than black paper, right? White is, well, white. That's all colors bouncing back to your eye. And black is absorbing all the energy it can; it's not letting any light bounce off to hit you in the eye. So it makes sense that white paper is also a better reflector of infrared light. Kinda cool.

Ultraviolet Light

If infrared light is on one side of ROY G. BIV, what's on the other? Another type of light called "ultraviolet" [ULL-truh-VIE-oh-lut]. In Latin "ultra" means above or "beyond." So ultraviolet light is the light beyond violet. Its frequency is higher than violet.

Ultraviolet light can do some pretty weird things. For one thing, ultraviolet rays from the sun can give pale people suntans and sunburns. Luckily, we have sunscreens to protect ourselves. Look on the side of the sunscreen bottle and you might see something about "**UV protection**." UV stands for ultraviolet light.

Iroy G. Bivu

So, let's talk about ROY G. BIV again and think about these wavelengths of light we can't quite see—infrared and ultraviolet. It's easy, the guy's name should obviously be IROY G. BIVU. Well, I admit I've never met anyone ever named Iroy. Nor have I ever seen the last name Bivu. I biv not; bivu? It's okay. Take a second and remember this: **IROY G. BIVU.**

So here's a diagram of all the types of light we've played with in this chapter: the frequencies we can see, and the two on either side. Just remember IROY G. BIVU.

Wavelength	Type of Radiation	Frequency
720 nanometers	Infrared	420 trillion Hertz
680 nm	Red	440 trillion Hz
620 nm	Orange	480 trillion Hz
570 nm	Yellow	530 trillion Hz
530 nm	Green	570 trillion Hz
470 nm	Blue	640 trillion Hz
450 nm	Indigo	670 trillion Hz
420 nm	Violet	710 trillion Hz
380 nm	Ultraviolet	790 trillion Hz

Hertz = one wave per second

IROY G. BIVU

Light Fantastic

Some things give off light when they're heated, like iron that glows in a blacksmith's forge. Some give off light when they have electricity sent through them, like the filament in a light bulb. And some chemical reactions produce light; that's how those glowing tubes you might have seen at concerts or fairs work. Here are a couple of other interesting ways we get light.

LASERS

You've probably heard of "lasers." The word laser actually came from the letters that start each word in this phrase: "**Light Amplification by Stimulated Emissions of Radiation**." What it means is that light can be amplified, or made more powerful, by giving it a boost with electricity. If light is amplified enough, it can melt metal! Other lasers don't cut, like the laser in a compact disk player. It just measures the depth of a tiny hole under the clear plastic of the CD, and that measurement tells the machine what sound to produce.

What happens in a laser, no matter how powerful or weak, is that the waves of light all end up moving together. Imagine our Light Wave Marching Band members all being exactly the same size and walking in exactly the same way and no one out of step at all. That's how light comes out of a laser.

Fluorescence

You're skating around the roller rink, and the manager turns off the main lamps. Suddenly, some of your clothes take on this weird glow, especially white cotton shirts and shoelaces and t-shirts with special paints on them. Where does this light come from? After all, there's no source of visible light around to reflect.

The manager has turned on what people sometimes call a "**black light**." That's not the best name for it since black means no light at all. Actually, "black lights" give off ultraviolet light; your eyes just can't see it. If ultraviolet light hits certain materials, it's sent back out as visible light. First the cotton or the t-shirt paint absorbs the ultraviolet. Then the cotton or paint gives the energy of the ultraviolet back off as light that we can see. It's weird the first time you see it. It's weird every time you see it, come to think of it.

When something absorbs an invisible light and gives it off as visible light, it's called "**fluorescence**" [FLOOR-ess-enss]. It's from the Latin word for "flow." The light seems to flow out of the glowing shirt.

Okay, what causes fluorescence? Well, to talk about this, we have to change how we think about light. Instead of talking about waves of light, let's talk about particles of light. Hold on, hold on, particles of light? Yes. Very small packets of energy that we scientists call "**photons**" [FO-tahnz]. These are the same photons that would make "photon torpedoes" on *Star Trek*. But hey, remember that *Star Trek* is science fiction—right now, there's no such thing as a photon torpedo, but there really are photons.

So what do photons have to do with anything? Okay, when photons hit something that fluoresces, like certain types of paint, the electrons in the paint jump up to slightly wilder energy level. They're still part of the same atom, but they're more "excited." Now, electrons don't stay in an excited state for a long time. They fall back down to their original energy level. When they do, the electrons give back another photon. That photon can hit our eyes as visible light. So, in the case of fluorescent paint, ultraviolet light photons hit the paint, and visible light photons are given off.

Fluorescent lights work much the same way. There's a gas inside the tube that gets jolted with electricity. The electrons jump up to higher energy levels, and then fall back down, giving off a visible light photons. So the light you see from a fluorescent light tube is light that's being given off by the glowing fluorescent paint inside the tube. Pretty wild.

Quantum Leap

Then you might wonder why we had to say that *separate* particles of light make the electrons jump. Well, it turns out that electrons only jump a certain amount. They either jump up or they don't. Then they either fall down or they don't. There is no in-between place, where electrons jump half way up or fall part way down—no middle ground. The smallest amount that electrons jump is called the "quantum." There's no "half quantum."

The quantum is the very smallest amount of energy that you can have in the universe. And that, by the way, is the amount of energy that we say is in a photon. The "quantum leap" is the very smallest change in energy that we can find in the universe. So it's goofy that when people talk about a quantum leap, they're often talking about a big change. On the other hand, a quantum leap is a big idea in understanding the nature of light.

See, here's an important thing about light that can be kind of confusing: light sometimes acts like a **wave** and other times it acts like a **particle**. It's wild! Actually, light always acts like light; waves and particles are ways we scientists have learned to think about it.

So we know about all the colors of light we can see, and that infrared and ultraviolet are on either side of the spectrum of visible light. But there are more forms of energy radiation on either side of infrared and ultraviolet. We'll talk about the rest of them in Chapter 7.

WILD!

ELECTRICITY & MAGNETISM

chapter six

Electricity and magnetism are wild things—they're the flow of pure of energy. Electricity and magnetism are also closely related—very closely. In fact, you can't have one without the other. That's a little bit hard to believe at first. What do refrigerator magnets have to do with light switches? In a word, everything. In fact, magnetism and electricity are so closely related that we scientists call this force by just one name:

Electromagnetism.

Every telephone, television, computer, video game, spark plug, light bulb, heat lamp, radio, flashlight, compass, hair dryer, airplane radar, walkie-talkie, movie projector, pocket calculator, clothes washer, copy machine, burglar alarm, and garbage disposal depends on our understanding of electromagnetism to work. So let's start with magnets.

Magnetism

Magnets either pull toward each other or push away from each other. Get a couple of refrigerator magnets and you can check this out for yourself. Usually refrigerator magnets are strong enough for you to use one magnet to push another around on a table top and never have them touch each other. Try it. When they push away, that's called **repelling**; when they pull together, it's **attracting**. This is the first rule of electromagnetism:

Opposite charges attract; like charges repel.

You might wonder how magnets can push and pull on each other, attracting and repelling. All magnets have "**force fields**" around them. You may have read about force fields in science-fiction stories. Some of those force fields are extremely powerful and kind of unbelievable, but magnetic force fields are real. Do you want to see one? Well, to tell you the truth, you can't. They're invisible. But you can see exactly where they must be.

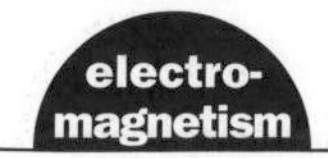

e x p e r i m e n t

Magnetic Force Field

Here's What You Need:

- *Beach or other big pile of sand*
- *Strong magnet*
- *Two plastic bags*
- *Sheet of paper*

Here's What You Do:

1. *Put the magnet in the plastic bag.*
2. *Drag the bag with the magnet inside through the sand. Particles will stick to where the magnet is touching the bag.*
3. *Turn the bag inside out around the magnet, and you can capture the magnetic particles. Shake them into the other plastic bag.*
4. *Put the magnet back into the first bag and start over. Gather about 3 mL (1/2 teaspoon) of magnetic particles.*
5. *Place a magnet under a sheet of paper on a table top.*
6. *Sprinkle the magnetic particles on top. How do they arrange themselves?*

Here's What You See:

You see the lines of the magnet's force field. Right there. You can imagine the "lines of force" coming up of the table and down into the table as well as around on the surface of the table. Try rotating the magnet, and moving the paper, to see where the magnetic field becomes too weak to move the particles around. Pretty cool.

Earth as a Magnet

Where do magnets come from? Actually, they turn up in nature. See, the Earth is mostly melted metal. We live on a thin shell that's cooled off. But a long way under our feet iron and other elements are churning around. It's so hot down there that all the metals and rocks stay molten. That lets them flow. And the flow

of the charged particles in the metals and rocks creates a magnetic field, turning the whole Earth into a big magnet.

The Earth's magnetic field gives certain rocks on the surface magnetic fields, too. If a rock that contains iron is located in a strong part of the Earth's magnetic field, it often gets magnetized. Once in a while, volcanoes erupt on the surface of the Earth, letting rocks that are mostly iron work their way to the surface. These rocks are sometimes magnetic and we call them, of all things, "**magnetite**" [MAGG-nih-tight]. That means "**magnet mineral**."

Compass

Ancient scientists found they could use magnetite to help make a compass. Now, a compass needle is a magnet that can move easily. It lines up with the Earth's magnetic field, which means one side points to where the magnetic field lines come out of the Earth. If you ever so carefully suspend a magnet on a thread so that it's free to rotate, it lines up with the Earth's magnetic field.

When a compass lines up with the Earth's magnetic field, it ends up pointing at "magnetic north," which is a point in the Arctic Circle close to, but not quite at, the North Pole. In other words, a compass points north. Knowing which way is north may not tell you exactly where you are, but that's all that explorers had to go by for hundreds of years.

You can make your own compass in about the same way ancient sailors did. Remember, all you need is a magnet that's free to move around easily.

Make Your Own Compass

Here's What You Need:

- *Magnet, like the type used to hold notes on refrigerators.*
- *Needle*
- *Cork, plastic packing "peanut," or a piece of sponge—something soft that floats.*
- *Glass of water*

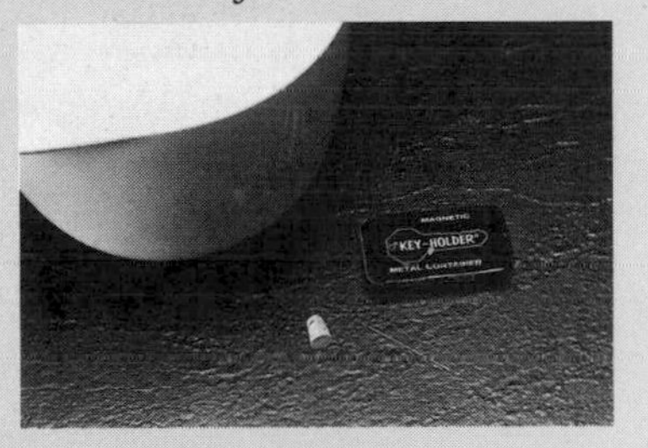

Here's What You Do:

1. *Turn the needle into a magnet. It's not that hard: Start by stroking it along the surface of the magnet. The more you rub it across the magnet, the more magnetized it will become. Try 50 strokes.*

2. *Carefully push the needle into the end of the cork or sponge.*
3. *Place the cork in the glass of water. Wait until it stops moving, and look at the way it points.*
4. *Turn the glass around. Where does the needle end up pointing?*

Here's What You See:

The needle points north—unless you have some more powerful magnetic field nearby. Carry your compass around the house and try to find magnetic fields. Try magnets, of course. And what about large appliances that are running on electricity? Hmmm.

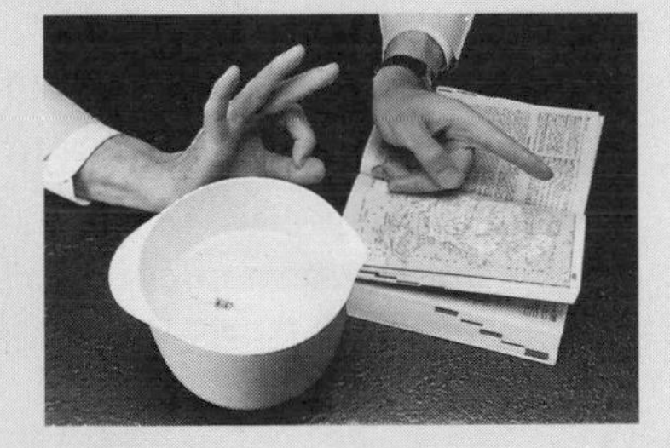

Compare your compass to a regular ship's compass, or the type you can put on a car dashboard. Many of those compasses work exactly like yours. They use a good strong magnet held by something that floats very evenly. And for the liquid they usually use something that won't freeze, like oil. The idea is exactly the same.

Iron

The key to the magnets we've discussed so far is that they all use the element iron. What's so special about iron?

Fe

Not this kind of iron.

There's something very special about iron's electrons. The electrons zipping around iron atoms can get lined up with magnetic fields and make a magnet. Once the electrons in iron are lined up, they pretty much stay lined up for a long time. Scientists (that's you) call the microscopic lined-up parts of iron "domains," like a queen's domain—the area she controls. So the domains in iron are controlled by magnetism. Iron's zipping electrons make its own magnetic field. Wild.

Very few elements can be magnetized:

- **iron**—*far and away the strongest and the most common.*
- **nickel**—*but the nickel in our five-cent pieces has been squeezed so much that its domains are distorted, and it's hardly magnetizable at all.*
- **cobalt**—*can make great magnets, especially when mixed with other unusual elements called the "rare earths."*

Try sticking a magnet to a steel appliance (like an oven) and to a stainless steel fork. Steel is magnetic, but stainless steel isn't. Yup. That's because stainless steel is about one-quarter nickel and **chromium** [KROH-mee-umm]. That's enough to make magnets not grab it. (Stainless steel cooking knives often contain enough iron to make magnets stick weakly.) By the way, shiny chrome car bumpers are almost all steel with just a thin coating of chromium. So magnets stick to the bumpers the same way they stick to painted steel doors on your refrigerator.

Electricity

Now let's switch to electricity. We all pretty much take electricity for granted nowadays. Everybody knows that electrical things like televisions or microwave ovens don't work unless they're plugged in or connected to a battery. But when electricity was discovered, no one had seen it work very much. There was a lot to figure out.

Of course, centuries ago people knew that electricity in some form existed because they knew about what we now call "**static electricity.**" That's Latin for electricity that "stays" in one place. Ancient Greek people knew about static electricity. Now, you can too.

On a dry day, you may have shuffled across a carpet and produced a shock as soon as you touched a metal doorknob or another person. That's static electricity: electrons build up on your body from the carpet, and then jump to something metal.

It might be weird to think about ancient Greeks shuffling around on carpets and giving each other shocks, but they might have. That's one way to start investigating electromagnetism.

experiment

Static Shock

Wait for a dry day. Often winter days are dryer than summer days (to find out why, you have to wait for Chapter 8). Put on shoes or socks, or both, and shuffle across a carpet. You'll build up a nice charge of static electricity on your body: extra electrons. Now what can you touch to make a shock?

Make a list of the things you can shock: metal, other people. And the things you can't shock: wood, glass, plastic. Already you're discovering something important about electricity. Some things are more electric than others.

Finding Loose Electrons

Here's What You Need:

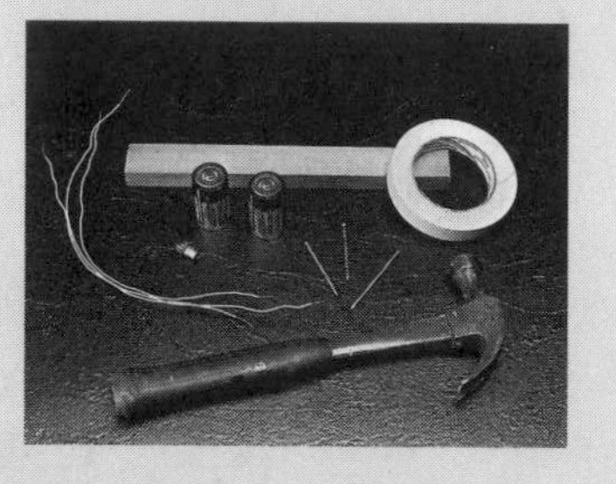

- *Flashlight bulb*
- *Type D battery*
- *3 lengths of wire, about 30 cm (1 foot) long*
- *Piece of wood*
- *Nails*
- *Hammer*
- *Objects to test: knife, fork, or spoon; plastic flying disc; glass; wood; paper clip; necktie; book; key; pot; or clay plate.*
- *Pencil lead from a #2 or #1 wooden pencil split longways*

Here's What You Do:

1. *Take the battery and the light bulb from the flashlight.*

2. *Lay the battery on the piece of wood, and hammer two nails into the board on either side of the battery. That will make the battery stay firm.*

3. *Wrap one wire around the nail at the top of the battery, and attach the other end of that wire to the light bulb.*

4. *Wrap another wire around the nail at the bottom of the battery.*

5. *Attach the third wire to the light bulb. When you're done, the set-up should look like this:*

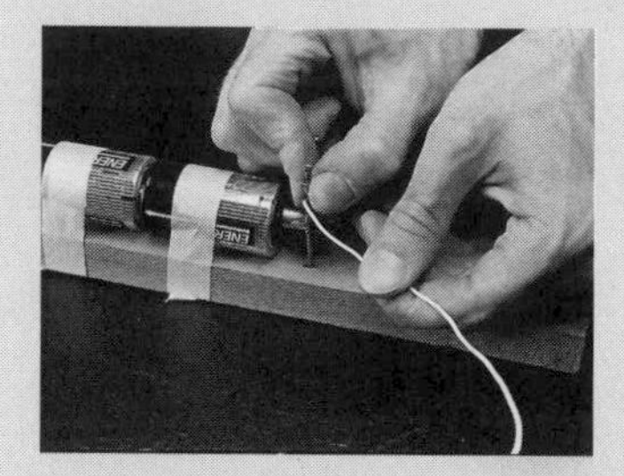

A. *Nails touching the top and bottom of the battery.*

B. *Wire running from the nail at the top of the battery to the bulb.*

C. *Wire running from bulb.*

D. *Wire running from the nail at the bottom of the battery.*

6. *Touch the two loose ends of the wires to the objects you find around the house. If something lights up the bulb when you touch the wires to it, put it in one pile. If the bulb stays dark, put it in the other pile. Can you see any patterns in what works?*

7. *Touch the wires to the ends of the pencil lead. Then move the wires closer together, until they're both touching the lead almost at the middle. How does the light bulb light up each time?*

Here's What You See:

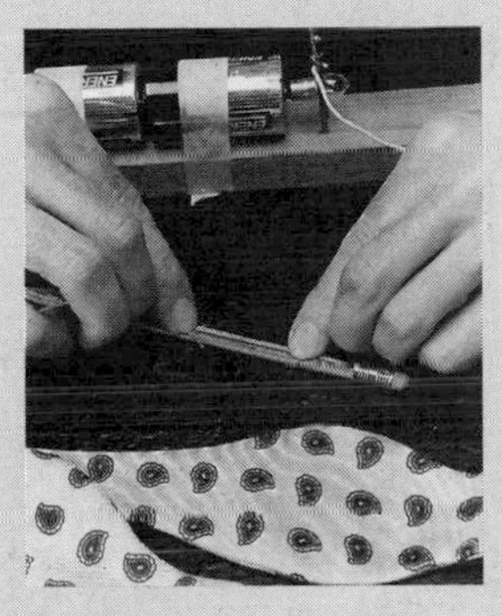

Some of the objects will make the bulb light up—most of these things are metal. The rest will be plastic, wood, paper, cloth, clay, and other types of matter. What's the difference? Everything has electrons, but in some materials electrons flow more easily than in others. Materials, like metal, that allow electrons to flow easily are called "**conductors**."

Is pencil lead a conductor? Pencil leads aren't really made of the element lead. They're a mixture of carbon and clay. Carbon by itself makes a nice black mark, but it's too soft to write with. So we add clay to make it harder. When clay dust and carbon are packed together, the mixture is hard enough to write with and black enough to see. A clay plate doesn't conduct electricity, but a pencil lead conducts a little. That means carbon makes pencil lead a conductor. Soft pencil leads usually have more carbon and conduct electricity better.

How well does a carbon-and-clay mix conduct electricity? The bulb lights up as it does with the metal objects, but the light doesn't burn quite as bright. And the closer the wires are to each other, the brighter the bulb. That shows that the carbon-and-clay conducts electricity all right, but not easily. Now, touch the wires to the pencil lead, the part that makes the mark, with the wires, first with wires at each end, then with the wires closer and closer together. You'll see that the light gets brighter as the wires get closer and closer together.

Electricity is moving electrons. They flow almost like water molecules in a river, so scientists (like you) call this an "**electrical current.**" But you need a loop of wire or other conductors to make electricity flow. Otherwise, the electrons stop. Scientists (like you) call this loop a "**circuit**," which comes from the same Latin word as "circle." So that's the next thing to remember about electromagnetism:

Electrical current flows around a circuit of conductors; if you break the circuit, electrons stop flowing.

Batteries

So where does the flow of electrons start? One source is chemical reactions. This is how batteries create electricity. See, some metals hold onto their electrons more loosely than others. If two different metals are put in an active chemical with a wire running between them, the metal that holds its electrons more loosely reacts with the chemical and sends electrons around the wire to the other metal. That creates an electrical current. When the metals' reaction with the chemical stops because they're all used up, the battery is dead. Most batteries use some type of acid for the active chemical.

e x p e r i m e n t

The Power of Money

Get a nickel and a penny that was minted before 1988. Pennies that old are made entirely of copper. Start by washing them with hot water and soap. (So many people handle coins that they can have a lot of germs on them.) After they're clean and dry, touch the nickel and penny to the tip of your tongue very close to each other. What do they taste like? Where both metals touch your tongue, you taste sweetness. That's because they're setting up a very small electrical flow in your saliva. That electricity stimulates your taste buds, sending your brain a "sweet" signal. But it all comes from the two metals.

WARNING!

Scientists in Benjamin Franklin's time had to make all their electricity from batteries. Now we use them when we want portable electrical power, or when we don't want to rely on electrical power plants.

Car batteries are usually what are called "lead-acid" batteries. Lead is a metal; the acid is usually "sulfuric" acid, which is very strong. So don't ever break open a battery; that acid can give you a very bad chemical burn.

WARNING!

You can make your own battery by setting up a chemical reaction using different kinds of metals and the acid in a lemon. No kidding. But we need two different metals.

WHOA!

Lemon Batteries

Here's What You Need:

- *3 fresh lemons*
- *Three two-inch pieces of half-inch copper tubing or two 15 cm (16 inch) pieces of bare copper wire from a hardware store. Or collect 15 clean pennies made before 1988, and clip them into three equal stacks with clothespins.*
- *Paper clips*

- *Aluminum foil*
- *Empty cereal box*
- *Scissors*
- *Liquid Crystal Display (LCD) clock, showing black numbers on a gray background.*

Here's What You Do:

1. *Cut the cereal box into strips, and wrap aluminum foil around them.*
2. *Stick one aluminum strip into each lemon with a paper clip on top.*
3. *Stick one piece of copper (or stack of pennies) into each lemon so that the copper and aluminum are close but not touching.*

4. *Attach wires like this:*
   - **A.** *aluminum in lemon #1 to copper in lemon #2*
   - **B.** *aluminum in lemon #2 to copper in lemon #3*
   - **C.** *aluminum in lemon #3 to copper in lemon #1*
   - **D.** *loose wires from aluminum in lemon #1 and copper in lemon #1*

5. *Now, hook the loose wires to the clock. Hook the wire from the copper to the plus side (+) of the clock. Now the whole thing should look like this:*

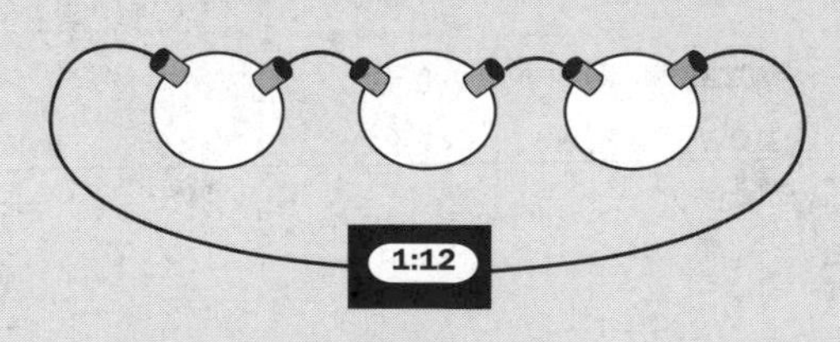

6. *Watch the clock. It should turn on and run for hours.*

Here's What You See:

Lemon batteries? Well, lemons have acid in them; that's why they make you pucker. Aluminum and copper are different metals, right? Why not? The way this is set up, the copper is always the positive "**electrode**" [ee-LEK-troad] and the aluminum is the negative electrode. Electrons flow from the aluminum to the copper.

How come we call one side of the battery "positive" and the other part "negative"? Well, that goes back to a wrong guess that Ben Franklin made in the 1700s. He thought that positively-charged particles moved from the copper to the aluminum. In fact, negatively-charged particles—electrons—move the other way. The clock runs either way you think about it.

BENJAMIN FRANKLIN

1706-1790. Experimented with static electricity.

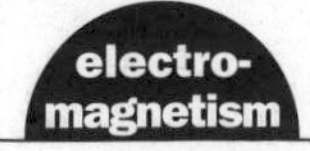

& The Quivering Compass the Queen

In 1820 a Danish guy named Christian Ørsted saw electricity turn a wire into a magnet—it even made a compass change direction. This was an extremely powerful discovery about electromagnetism: **An electrical current sets up its own magnetic field.**

MICHAEL FARADAY

1791-1867. Studied how electrical currents create magnetic fields.

About a year later, an English guy named Michael Faraday showed his Queen how an electrical current could make a compass needle swing around. He said something like: "Look, your highness. I can make that metal needle move by connecting and disconnecting these wires." Do you know what she said? Well, actually no one knows for sure. But the story goes that she said, "Of what use is it?"

I like to think that Michael Faraday was just about ready to explode when he heard the Queen's response. He might have been thinking: "HEY! QUEEN! Would you look, for crying out loud? I'm not touching the needle. No one is touching the needle! I am moving the needle with some invisible force traveling through the air. Come on, Queen! This is a great discovery! This is important! This is cool!"

GAH!

But instead, all Faraday said was, "**Madam, of what use is a new born babe**?" You know, a new baby isn't very useful. They can't carry groceries. They can't do the dishes. They're a lot of doggone work. But a baby grows up. And some of them get to be scientists, win Nobel prizes, get rich, and so on. So Faraday realized that he was on to something. Eventually that invisible force would grow up, too. And it has! Every electric light, every computer, every telephone, every microwave oven, every television, every video game—everything electrical is, in a way, a product of this discovery. Not bad.

Electric Magnet

Now we start to put everything together: magnets and electric current. Actually, magnets and electricity have always been interconnected.

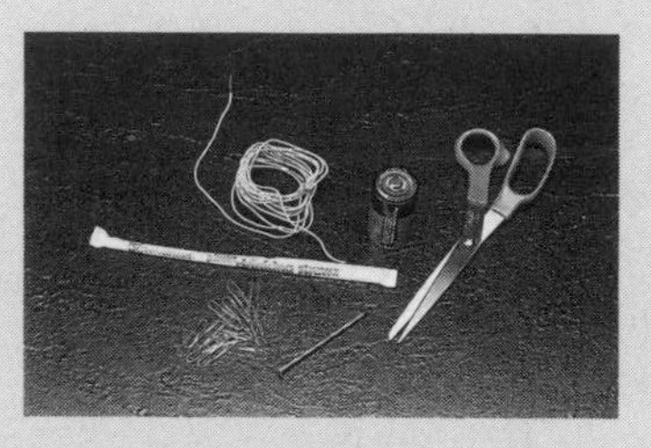

Here's What You Need:

- *Drinking straw; the thicker and stiffer it is, the better.*
- *Steel nail that will slide down into the straw. Test whether the nail is steel by checking if it sticks to a magnet.*
- *Piece of insulated electrical wire (coated with rubber or plastic)*
- *Type D battery*
- *Steel objects. Try paper clips, bike frames, bike chains, and the steel parts of a ball point pen.*

Here's What You Do:

1. *Put the nail in the straw to hold it stiff, then wrap the straw with a long piece of electrical wire.*
2. *Attach the ends of the wire to the battery.*
3. *Try to pick up some steel things with the coil of wire. Now that there's an electrical current running through it, the wire has a magnetic field.*
4. *Take the nail out of the straw. Watch what happens when you try to pick up steel things now.*

Here's What You See:

An ordinary non-magnetic coil of wire turns into a magnet when electricity runs through it. It's even stronger when there's iron inside the coil. The magnetic domains in the iron boost the magnetic field big time. It's a little weird. If you use almost anything other than iron (or steel, which is mostly iron), the magnet is pretty weak. But if you build an electromagnet in the right way, it's extremely powerful.

Electric Motor

WILD!

Now it's time to put electromagnetic forces to work. To power something with electricity, we want to convert the electromagnetic energy into movement. This is how electric motors work, such as those in an electric mixer. You can build an electric motor. Give it a shot.

Motoring

Here's What You Need:

- *Strong magnet. Use a magnetic "Hidden Key" box. Some large refrigerator magnets will work.*
- *Cork*
- *2 safety pins*
- *A small piece of corrugated cardboard like the flap of a shipping box. It's something to stick the pins in.*
- *A nail file or very small knife*
- *Type D battery*
- *4 feet of thin "enameled" solid wire. Sometimes this wire is called "magnet wire." You get it at an electronics supply store.*
- *Adhesive tape*

Here's What You Do:

1. *Place the magnet on the piece of cardboard.*
2. *Stick the pins into the cardboard so that they are wide enough for the cork to fit in between plus a little, say 1 cm (1/2 inch). Tape them in place.*
3. *Wrap thin wire around the ends of the cork—the long way. You're*

forming a tight coil of wire around the lightweight cork. Bring the ends of the wire out as straight as you can.

4. *Now, gently file or scrape the insulation off the wire so that half of it shows the shiny metal underneath. Make both sides shiny on the same half.*
5. *Gently place the cork between the safety pins with the straight wires running through the round loops of the safety pins. The cork should be loose enough to turn.*
6. *Hook the pins (not the straight wires on the cork) to the battery with some extra wire. Give the cork a slight spin to get it started, and it will run with energy from the battery. You've made an electric motor!*

Here's What You See:

The coil forms an electromagnet every time the bare wire ends are touching the safety pins. It repels or attracts the magnet on the cardboard. Because it's never quite in the center long enough to stick, the magnetic fields push the cork so that it twists or spins. Not bad. This is basically how all electric motors work!

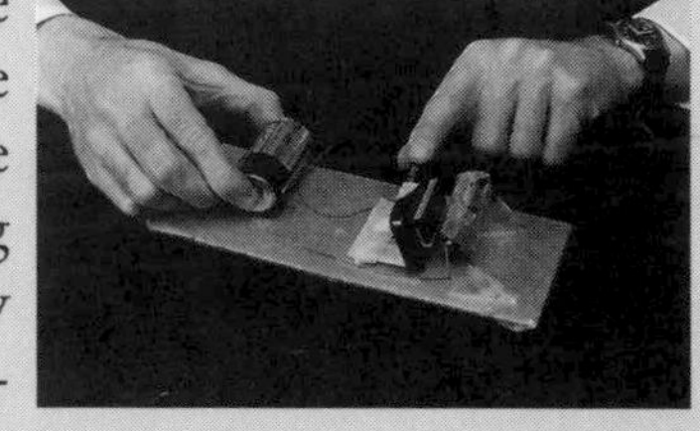

Conductors moving near a magnetic field create an electrical current.

Electromagnetism at Work

The key to the connection between electricity and magnetism is that things have to be moving. Electricity is the flow of moving electrons. So for a magnet to make electricity, either the magnet or the electrical wire has to be moving. Almost all the electricity in your house is made by very large machines called "generators" or "dynamos" (DYE-nah-mohz) that have coils of wire moving near magnets.

First, we need a source of energy. Let's imagine it's a windmill. The wind turns the windmill, which turns a big coil of wire in the electric power plant. (Other types of power plants turn the wire coils in other ways.) The coil of wire is near a magnetic field, so as soon as the coil turns, an electric current runs through it.

The power plant is connected by more wires to your house. So whenever you flip a switch, you complete a circuit with the wires to the power plant. That means the electricity generated at the plant can flow through the wires in your house.

Some electricity goes through a light bulb, sending out light and infrared energy (heat). Or it goes into a CD player, making the disks move, the display panel light up, the laser beam inside turn on, and the speakers to send out sound waves. Or your electric alarm clock—you get the idea. Everything that runs off electric sockets depends on a coil of wire turning near a magnet somewhere.

Whenever a current of electrons is moving, all the electrons flowing along together, we get magnetism. And, whenever magnets are moving around things that can carry electricity, like wires, we get electricity. This connection allows us to create and control electricity, which makes our whole world go.

RADIO, TV & MORE

chapter seven

You see TV and hear radio nearly every day, but what exactly are they? Radio and television are sounds and pictures that are changed into electricity and sent out through the air or space. We can pick up a tiny part of that electricity with a radio or television set, which lets us see and hear the sounds and images. Some kinds of radio even go through the oceans. How can this work? Let's start out by talking about electricity.

Go with the Flow

Electricity is the flow of the particles we call electrons, remember? One way of thinking about electricity in wires is to imagine a water hose that's full of water and connected to a spigot. As soon as you turn on the spigot, water comes out the other end. The water coming out of the faucet almost instantly pushes water out the other end of the full hose. Each molecule of water pushes on the next. Electric currents work the same way, except electrons push on each other instead of water molecules.

Very soon after electric currents were discovered, people realized that they could be used to send **information**. You might wonder what I mean by information. That's words, numbers, pictures, sounds. The key is that those things are converted into a code that both the sender and receiver understand.

e x p e r i m e n t

Flow of Information

If you have a friend, a garden hose, and a house, and there's no water shortage, try sending information with the flow of water. You stand at the water spigot where the hose is attached, and your friend carries the other end of the hose around the corner of the house. In other words, you can't see each other, but this hose runs between you.

Run the water for a few moments so the hose is full. Then turn on the water in bursts. Some water will come out the far end every time you turn on the spigot. You can send information to your friend with the water flow. Use squirts of water as signals. Make up a code. One squirt can mean, "Yell the word 'Lizard'!" Two squirts can mean, "Throw a ball out where I can see it." And maybe three squirts means, "Let's get some ice cream." (You might want to practice that last one pretty often.) Does your friend get your messages?

So you can signal your friend with the flow of water. What if your buddy wants to signal you back? You could do that if you had two hoses and two faucets. It would be like a water walkie-talkie. If the hoses were set up so water flowed around and around, it would be a circuit—just like an electrical circuit. With a circuit, both you and your friend could send information to each other.

The Telegraph

In 1837 a painter named Samuel F. B. Morse thought scientifically about electrical circuits and designed a way to send information along a wire. He invented what came to be called the "telegraph" [TELL-eh-graff], which means "writing from a long way away." It could send messages as far as the wire was strung: a lot farther than you can see, and a lot farther than you'd want to run a garden hose.

The idea behind sending information with a telegraph is to make signals with a

SAMUEL F. B. MORSE

1791-1872. Invented the telegraph.

small metal magnetic lever. Every time you push the lever down, it completes an electric circuit, and a lever in another city is pulled down by magnetism. That makes what's called a "click." There are two kinds of clicks: long (dashes) and short (dots). Normally, a dash is about three dots long. With one click being three times the length of the other, you seldom confuse the two. Because he was in the business right at the start, Morse came up with the standard information code people are still using today. He assigned particular sets of clicks for all of our letters and numerals. We call this, naturally, the "Morse Code."

You don't have to use an electric telegraph to send Morse Code messages; you can make these signals with light, sound, squirts of water, or smoke. It takes practice to get where you can receive them and understand them. It's pretty cool, though.

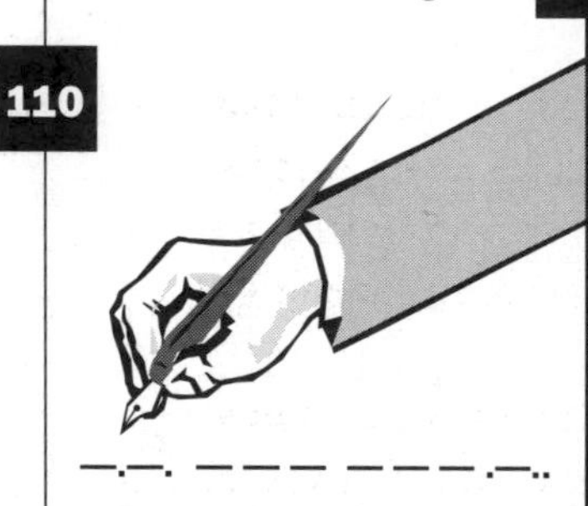

MORSE CODE			
A .—	J .— —	S ...	1 .— —
B —...	K —.—	T —	2 ..— —
C —.—.	L .—..	U ..—	3 ...—
D —..	M —	V ...—	4 ....—
E .	N —.	W .—	5 .....
F ..—.	O — —	X —..—	6 —....
G — .	P .— .	Y —.—	7 — ...
H	Q — .—	Z — ..	8 — ..
I ..	R .—.	0 — — —	9 — — .

The Telephone

The next jump was the telephone, which means "voice from a long way away." Instead of simple clicks, the telephone sends our voices over wires. Actually, it doesn't exactly send our voices. Instead, the telephone converts our voices into electrical signals, which travel over the wires to another telephone. Then the second telephone converts those signals back into the sound of your voice.

The telephone was such a great step forward because sound contains a lot more information than letters and numerals. How long does it take you to recognize a friend's voice on the telephone? Not very long—about a quarter of a second. It would take a lot longer if your friend had to spell out his or her name in Morse Code.

You can even tell right away whether your friend is excited or sad, in a busy outdoor place or a quiet room. How long would it take you to find all that out if your friend had to send you Morse Code? Or what if your friend wanted to play you a new tune for the kazoo? On the telephone the sound would come right to you. But there's no Morse Code for music. See? (Or rather, hear?) There is a tremendous amount of "information" in sound, and it travels much faster than the ol' code.

Mister Microphone

For telephones to work, they need a microphone. That word means "very small sound," because some microphones are used to make very small sounds bigger and easier to hear. But basically microphones convert the energy of sound into the energy of electricity.

You may have seen a microphone, or even spoken into one. How does it work? To start with, hold your hand in front of your mouth while you talk. You can feel the air move as you talk. That moving air is the sound of your voice. Pressure waves are set up in the air by your voice, or by a clarinet, or by whatever is making the sound.

When the moving air hits a microphone, it wiggles some very small part inside the microphone. Some microphones use tiny magnets; some use a tiny metal plate charged up with extra electrons; and some use a special type of mineral crystal that affects the way electricity flows. Anyway, whatever type of microphone is used, it converts the very small movements of the air into very small electrical signals.

Look, Ma, No Wires

By now, you might be saying, "Hey, hey! I thought we were going to talk about radio and television." Well, we are, in a way, because we're talking about sending information. That's what telegraph and telephones do, and it's also what radios and televisions do.

If you try sending Morse Code, you see that you just can't send it very fast (not nearly as fast as we talk and not nearly fast enough to show a whole scene). Telephones are faster and can send more information, like the tone of our voices, but they still need wires. You couldn't telephone a ship or an airplane. So, okay, what do we do?

We hurry. We send out signals at the **speed of light.** You may remember that the speed of light is very fast indeed. It's fast enough to go from Los Angeles to New York in about 2/100 of a second. And we send signals **through space**, not on wires. Light, infrared, and ultraviolet waves don't need any wires, remember. Neither do radio waves.

The Radio

Nowadays, most people give credit for building the first successful radio to an Italian inventor named Gugliemo Marconi [Goo-LYEL-mo Mar-KOH-nee]. (Hey, it's Italian.) He built on the work of other scientists. In 1895, he sent telegraph signals—dots and dashes—through the air. Marconi realized that electricity not only can travel through wires, but it also can travel through space. So right away his invention became known as "the wireless." (Now, a radio is full of electric circuits—it's full of wires, okay? But because there are no wires *between* radios, it was called a "wireless.")

GUGLIEMO MARCONI

1874–1937. Invented radio.

You probably have heard of the great ship Titanic. *It was thought to be unsinkable. That turned out to be wrong. The* Titanic *sank in 1912. A lot of people disappeared, but a lot of other people were saved because the ship was new and fancy, and it had a radio. The radio operator called other ships to come rescue the people in lifeboats.*

After the Titanic *accident, all ships were required by law to have radios. Can you imagine a time when giant ships would not have radios? Can you imagine landing an airplane without a radio? Now, having a radio seems like the most obvious thing in the world—why would you even need a law? But radio was new, and not everyone saw how useful it could be.*

Radio Waves

Okay, so what did Marconi do that was so cool? Ah hah. He knew that an electrical current sets up a magnetic field. Then he realized that if he made that electrical current go back and forth, switching very quickly, it would set up **waves** in that magnetic field. And then he realized that those waves were just like light waves except longer.

This type of wave is called "**electromagnetic**" energy [Ee-LEK-troh-mag-NET-ick]. That means it comes from the electromagnetic force. Electromagnetic energy goes out in straight lines. We say it "**radiates**" [RAY-dee-aits]. That's from the Latin word for "ray," like a ray-gun or the Sun's rays. Scientists (like you and me) like to say that light and radio waves are forms of "electromagnetic radiation" [RAY-dee-ay-shun]. It's cool.

Radio waves are one type of this electromagnetic radiation; light is another. Radio waves are just like light waves, only they have a smaller frequency—much smaller. That means they have a longer wavelength—much longer. Marconi figured that out by making electricity go back and forth through a wire. Then, by walking around his lab with another wire, he could find when the electricity was strongest and weakest. He found the wavelength of radio waves right there in the room he was standing in.

How long are radio waves? For a typical AM radio station that you might listen to a baseball game on, they're about 300 m (300 yards) long. It sounds like a long way, doesn't it? Remember, though, that these waves are going the same speed as light—that good ol' speed of light. So if you're standing somewhere thinking about these waves, it only takes about one millionth of a second for the whole wave to go past you!

One way of thinking of an electric field is to imagine a spray nozzle, like one you might use to wash your car, spraying a very fine mist into the air. If you're standing near the nozzle, the mist is thick, and you might get pretty wet. If you're standing farther away from the nozzle, the mist will be more spread out. Electric fields are a little bit like this. The closer you are to the source of the field, like the antenna on a radio broadcast tower, the stronger the field. Then when you're farther away, like maybe out of town, the field is more spread out. Notice that even though the field of mist has different strengths depending where you are, the mist is still moving away from the spray nozzle. And so it is with an electric field. The intensity or strength of field differs as you move around in it, but it's still there, and it's pretty much moving or tending to go away from its source.

Oscillating

Here's what Marconi and these other guys and gals in the 1890s realized: In order to have an electric field that you can send through the air (a radio wave), you have to have electricity that changes very rapidly, going one way and then very quickly going the other way. We call this "oscillating" [AH-sill-layt-ing] quickly. The word oscillate comes from the Latin word to "swing."

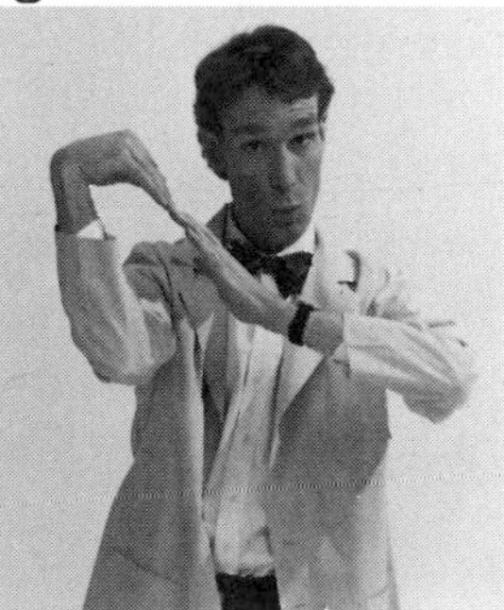

If you think about it, you can see that when something is oscillating, it's moving like a wave. If you're sitting in a boat, you feel the waves go past you and under you. But you don't move along with them very much; instead, you move up and down. You oscillate up and down as a result of the wave going past you. You can make your own oscillator and make your own waves—and not in water. Try this:

e x p e r i m e n t

The Wave

Here's What You Need:

- *Floor*
- *Big long spring, or a piece of thick rope*
- *Thumb tack*

Here's What You Do:

1. *Clear some space near the wall of a room where the floor is bare, or where there's no carpet.*
2. *Secure one end of the spring or rope to the wall, or the edge of a door, with a thumb tack or on a hook. Door stops are ideal.*
3. *Kneel so the spring or rope is laid out pretty straight, and take hold of the spring or rope. Then swing your arm back and forth.*

Here's What You See:

You will set up waves in the spring. The spring will be oscillating. Wave your arm back and forth quickly; that increases the frequency of the waves in the spring. Notice that their wavelength decreases—the waves get shorter. Then wave your arm slowly but steadily. The frequency goes down, and the wavelength gets longer.

Catching the Wave

Marconi realized that if electricity is oscillating fast enough, energy will go into a wire like an antenna and right off into space or into the air. Then, if things are set up right, you can use another antenna to capture just a tiny amount of that energy in the air and get information from it to understand the original signal. That's

what the "wireless" radio is all about. That energy is what we now call a "radio wave."

Now, you've seen various types of antennas all over the place. There's one on almost every car you see. There's usually one on your television. Radios have one. An antenna sometimes has a special shape, but basically it's just a piece of wire, or some other metal, that's connected to an electric oscillator and is free to broadcast or to receive electromagnetic signals out of or into the air. Antennas are electrical conductors, because they have to carry electrical energy.

In the earliest days of radio, people turned the oscillator on and off, or connected and disconnected the antenna, to create pulses of electrical energy. Those pulses could be made long or short. That's right. They could be used to send and to receive Morse Code. Pretty cool.

COOL!

The next big idea to improve the radio was discovered mainly by a guy named Reginald Fessenden. He was born in Canada and was working in the U.S. He realized that there was a way to use all this wild up and down motion of the electromagnetic radio wave to carry more information. His idea was to change the size of the waves as they were created. Rather than just turning the waves on or off, he figured a way to make them larger and smaller while they were being sent. This is called "**modulation**" [mahd-yoo-LAY-shun], from the Latin word to "measure."

REGINALD FESSENDEN

1866-1932.
Invented AM radio.

By modulating the amplitude (AM-plih-tood), or "bigness," of the radio wave, Fessenden was able to hook up a microphone to a radio antenna and convert a voice into radio waves. The small signals from the microphone control the size of the waves going out of the oscillator that is used to make the radio waves. Then the receiver could catch the waves and convert the changes in modulation back into a voice. This is what we call "AM radio," for "**amplitude modulation**": it's really "bigness modulation."

So let's imagine you're listening to a basketball game on AM station 1250. That means the station is sending out radio waves at a particular frequency: 1,250,000 waves per second. Some of those waves are bigger than others, which allows them to carry complex information. The information, like the voice of an announcer at a basketball game, is handled like this:

- *The announcer speaks, creating sound waves.*

- *The microphone converts the sound into electrical current.*

WOW!

- *The oscillator and antenna convert those electrical signals into radio waves.*

- *Another antenna catches the radio waves and converts them back to electrical current.*

- *The radio's speaker converts the electrical signals into sound.*

There are some pretty wild things that happen with AM radio signals.

Ionosphere

Get your parents' permission to stay up late, and get an AM radio. Adjust the tuning knob or buttons on the radio so that you're tuned between stations in your home town or area. On a good night, you can hear all kinds of stations that you can't hear normally. You might even hear people or computers sending Morse Code.

Those signals bounce back to your radio from a layer of the sky very high up. The Sun's energy makes this layer of the atmosphere electrically charged a little bit. Radio waves at the frequencies we use to send AM radio bounce off this level of the sky—off the "ionosphere" [EYE-ahn-nuss-feer]. During the day, the ionosphere isn't the right temperature to make radio waves bounce off of it. At night it is. It's pretty wild all the stuff you can pick up. The better your AM radio, and the better the weather, the more stations you'll be able to bring in.

Turning AM Radio Around

Get a small radio, and set it to an AM station that you can hear only moderately. Then rotate the radio: slowly turn it around so you can see each side in turn. Is the sound clearer or louder in any particular positions?

An AM radio receives the signal best when it's facing toward or away from the antenna sending out that signal. That's because the body of the radio works with its own receiving antenna to catch more waves. This experiment works best when you're pretty far away from radio stations because otherwise radio waves can get bounced around and come from several directions at once.

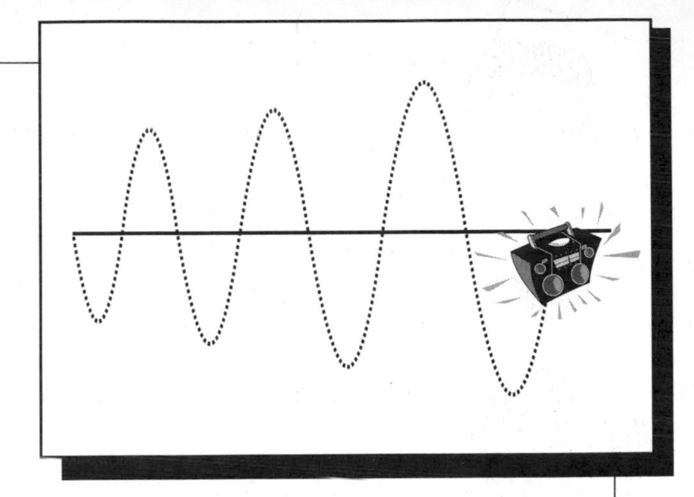

FM Radio

Now let's talk about the other kind of radio station, FM. That stands for "**frequency modulated**" radio. What do you think that means? Well, it means we modulate the frequency. Instead of the bigness of the waves, we change their wavelength. But the frequency is what makes one station different from another; it's what makes "89.5 KNHC" different from "107.7 KNDD." Hang on, hang on! So how can we modulate, or change, the frequency without losing our place on the radio dial?

Don't worry, we don't change them very much—hardly at all, actually. Let's take that second station that I mentioned just now. It is assigned the frequency of 107.7 Megahertz. A "Hertz" means one wave per second, and "mega" means "million." So 107.7 KNDD sends out 107,700,000 radio waves every second. That's pretty doggone fast.

Actually, for FM stations it turns out that 107.7 doesn't mean exactly 107.7; it means *around* 107.7. Those stations vary the waves' frequency up to around 18,000 waves per second. FM radios are set up to receive the whole signal and demodulate it without losing their place. Those 18,000 waves are plenty high enough to make music and people sound very good, but it's only about 1/5000 of the broadcast signal. There's not much difference between 107,709,000 and 107,691,000. To leave room for that modulation, the frequencies of FM stations are farther apart than those of AM stations.

experiment

Standing in the Way

For a typical FM radio station, radio waves are about 3 m (10 feet) long. That's about the size of a room. Tune your radio to an FM station that doesn't come in very strongly. Then walk around your room until you find a place where the sound gets crackly. When you stand in that place, your body is absorbing some of the radio waves.

That's pretty much how radio works. Sound is made into electrical waves, sent through space, and made back into sound waves. Well, television must just do the same thing with light and sound. That's a lot more information.

Television

Like lots of great inven–tions, television doesn't have just one inventor. A television has so many pieces and circuits that it would have been hard for just one person to come up with the whole system. Even though the idea of television isn't that complicated (people began to think about it as soon as they saw movies and heard the radio), getting television actually to work well took years. Here we go. . .

There are three things you have to do to send TV pictures and sound all around the world. It's just like radio, but you're working with light energy as well as sound energy. First, you make the light and sounds into electrical signals; next, you send the signals; finally, you make the electromagnetic waves back into pictures and sounds.

OOH, AHH!

Now, think about it for a moment, and realize how much information you need to have to get a good picture. You need a lot. If we think of it in terms of our good buddy Morse Code, we can begin to see just how many dots and dashes we'd need—a lot. Try this:

What's in a Picture?

Look at this picture of me with a magnifying glass. (You can also look at a black and white newspaper photograph.) You can see the picture is really a lot of very, very small dots, set up in regular lines. When the dots are big and close together, the picture is dark. When they're farther apart, the picture looks light. Your eyes take all the dots in, and your brain recognizes the pattern they create as something familiar enough to you for you to "connect the dots" and understand what the picture shows.

Each dot's size and position is a piece of information. Strangely enough, the farther the picture is from you, the easier it is for your brain to figure out. It's because the dots end up blurring together as they get farther from your eyes.

See if you can create your own picture just using dots you make with the point of a pencil. It can take a long time. Don't feel you have to finish it all at once. Here's a couple of tips, though. Plan your dots pretty close together. Also, try making your picture without too many dots and then put it across the room.

You can barely recognize me, right?

Moving Pictures

Now, let's say you want a picture to move (like it does on television). How would you do that? Ah hah, there's a trick—a trick about our eyes. It turns out that our eyes hold on to pictures for just a few instants even after the picture has gone away. Scientists like you call this "**persistence of vision**."

For most people, an image or picture stays in our brain for about 1/16 of a second. That means if I show you a picture, I have 1/16 of a second to flip another picture into its place, and your brain connects the two pictures as if one has just turned into another. If I show you 16 or more pictures in one second, it looks like something moved. Pretty weird thing about our eyes.

Movies you see in the theater take advantage of our persistence of vision. Movies are actually not a continuous moving picture. Instead, they are made up individual frames of film all strung together. Each frame is slightly different from the last. When someone moves his or her hand in the movies, his or her motion is being captured on individual pictures each slightly off from the one before. Because our eyes hold on to each image for a moment, by the time the next image comes on, the old image is just fading away. The result that we see is a continuous moving picture!

Early movies had sixteen pictures every second, just enough for most people's eyes to be fooled. But some people with particularly sensitive eyes could see the frames go by. Nowadays movies have 24 frames going by every second, and almost no one can tell. But when very old movies are played at this faster speed, the motion looks speeded-up and jerky.

So what does this have to do with television? Well, everything. Television, just like movies on film, takes advantage of our persistence of vision. In the U.S. televisions show 30 images every second. Our eyes put all those images together to create smooth, flowing movement.

Television Blinks

Here's What You Need:

- *Long thin pencil*
- *Television set*
- *Lamp with an incandescent light bulb*
- *Lamp with a fluorescent or neon light bulb*

Here's What You Do:

1. *Turn on the television and sit back.*
2. *Wave the pencil in front of the screen quickly.*
3. *Wave the pencil the same way in front of an incandescent light bulb.*
4. *Now try waving the pencil in front of a fluorescent light bulb.*

Here's What You See:

In front of the TV, you can see that the waving pencil seems to have several images. It's almost as though you were holding a sort of pencil fan. It's highly weird.

That sight is caused by the picture on the television being created and turned off twice for every frame of pictures. The light coming off the TV is blinking 60 times a second—twice for each image it shows! Because the pencil is such a distinct dark shape, you can see a lot of those blinks.

When you wave the pencil in front of a regular light bulb, you won't see any fan images. Now, wave it in front of neon or fluorescent bulb. You'll see the multiple pencils again. It's because those types of lights are blinking on and off 60 times a second too, just like the TV.

WHOA!

The blinking of lights comes from the power stations, where electricity is generated. It's made in a way that turns the electricity off and on again constantly—at 60 Hertz, or blinks per second, in the U.S. It works just fine.

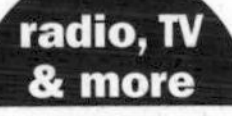

Collect the Dots

How many dots do you see in one second of television? There are usually more than 300,000 dots in each picture. And there are 30 pictures each second. Gee, that's almost ten million dots a second. Whew! Fortunately, we can control this information so quickly because we are working with energy carrying the information at the speed of light.

So okay, let's talk about making light into electrical signals because that's the key to making a television picture. It's the first step if we plan to beam these sounds and images around like radio waves.

You may have heard of an "electric eye" or "solar cell." These are names for the same kind of device. I guess the best name for them is "**photoelectric** [FOH-toh-ee-LEK-trick] **cells**," or "photocells" for short. Photoelectric cells convert the energy of photons into the energy of moving electrons. In regular words, they make light into electricity. Now, the amount of electricity the cells make is pretty small, and it's not like they "see" images. It's just that if enough light hits them, they make a small amount of electricity.

Photocell

Many times when you go into a store, you hear a chime, bell, or buzzer go off with the opening of the door. A light bulb is set up to beam light on the photocell. When the door opens and blocks the beam, the photocell senses that it's not being hit by photons. That changes the electric circuit, which sets off the chime. Okay, that seems like it would work.

Check out how this works next time you're in a store that has a photocell chime on its door. Start by asking the person working in the store if you can try this. Constant chiming can make you one unpopular scientist in a place of business. If you're nice, people will be nice.

Find the source of light for the photocell. Then move your hand in and out of the beam of light that shines on the small photocell. You can make the buzzer go off and on. It's pretty cool because you're not touching a switch, like a light switch or doorbell button. You're just interrupting a beam of light. Photocells are pretty cool.

In addition to photocells, there are other devices that change the way electricity passes through them with how much light hits them. These devices are called "**photoconductors**" [FOH-toh-kunn-DUCK-terz]. Just like photocells, these devices let light control electricity. That's the key to TV.

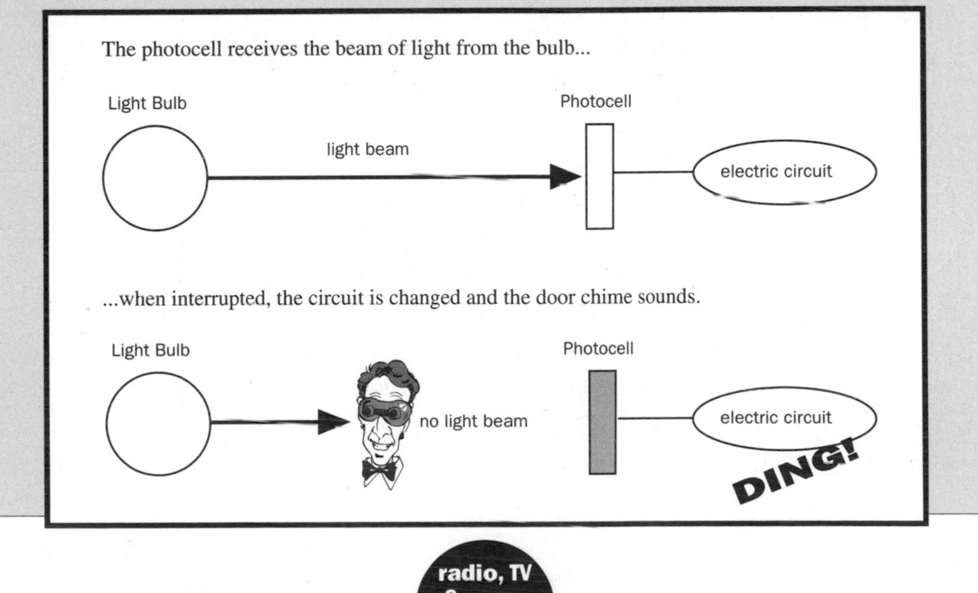

Inside a TV Camera

If you're thinking along with me here, we just need a way to get the picture that we're trying to put on television onto some of this photoconductive material, and then to get some sort of electrical signal off of it. The idea is amazing. It's like having many many photocells set up in a pattern that's about the same shape as a TV picture, and then getting the electricity off of them in the same pattern as the light that's hitting the cells.

Now here's how an image appears on your television: light comes through a lens in the front of a camera at the television studio, just like light through one side of a pair of eyeglasses; the light hits some photoconducting material and gets converted into electricity; then the electricity is sent where we want.

The big trick—the big idea that makes television possible—is called "**scanning**." To scan something means to look at it thoroughly. In a television camera, a photo conducting plate is scanned with a beam of moving electrons! The image—the newscaster or rock video—is converted into an electrical signal just a little bit at a time. By scanning the image, we are able to put the tiny bits of electricity in sequence, almost like dots. When we send all the information to a TV set, it shows those dots and makes them look like those images are right in our living rooms.

After we capture the picture with a camera tube, we still need electron beams to create the picture on TV at home or on a computer screen. You might wonder where one goes to get a beam of moving electrons. Well, scientists have been making these electron beams for years, and here's how it's done. The electrons are heated on a thin wire called a "**filament**" [FILL-uh-ment], from the Latin word for "thread." The heated electrons are loose enough to be pulled off the wire by an electric field in a nice straight steady beam. This set up is usually called an "**electron gun**."

Now, here's the next weird part: Just as a moving magnet can make electricity flow, a changing magnetic field can make this straight beam of electrons go off in funny directions. We use coils of wire to make the changing magnetic fields. If you change the magnetic fields fast enough and in just the right way, you can make the beam "scan" the inside of the TV plate. The beam goes from right to left very fast. Then it jumps up a little, and goes from right to left again, until it has scanned the whole picture.

I've got an electron gun and I'm not afraid to use it!

When we get the picture information to our television set, we can make the same picture the camera tube made by scanning it back out with a "picture tube." The picture tube spreads the scanned-in information from the camera tube back out on the glass screen. The screen is coated with what are called "phosphors" [FAHSS-ferz]. Phosphors are chemicals that glow when electrons hit them.

Now, I said before that the key to images appearing on television at first was "scanning." That's making a beam of electrons go back and forth across the back of the screen, just like spraying a hose when you're watering the lawn. The beams have to go precisely along lines of dots—the phosphors. Today in the U.S., 525 lines of phosphors make up a TV screen.

*Okay now, every time the electron beams sweep or scan over the back of the screen, they light up the phosphors on one row or line of the screen. Scientists figured out a trick: We can make the beams light up **every other line** on the screen, skipping one line with every pass. Then we have the beams go back and light up the lines we skipped over the first time. Because of our persistence of vision, our eyes don't notice the skipping and going back. It all happens too fast. Pretty cool, huh? Televisions in the U.S. are scanned 60 times a second. That adds up to 30 frames a second. On a typical 25-inch*

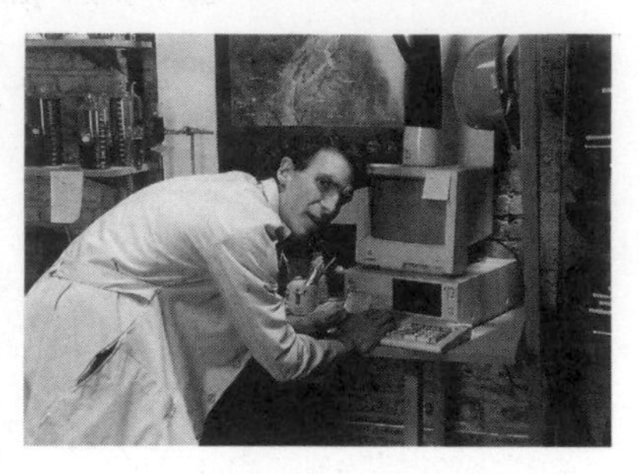

TV set the electron beams sweep through about a mile of lines on the screen every second! That's just amazing.

Here's something else cool: There's a little extra space in a TV signal. Even though there are 525 lines available, our TV sets show only about 490 of those lines. That means TV stations can use the extra lines to send special signals. For instance, some TV stations send the sound in a second language, like Spanish. Often shows are sent out with the spoken words printed at the same time right on the screen, for viewers with impaired hearing ("closed captioned"). You need special equipment to decode those signals, but they're out there. Think about that. You can send a show's words in another language completely separate from the main show, and most people can't even notice it! There's that much information in a few lines of TV screen scanning.

Color

Television in the United States was originally designed to be black and white—no color. That was in the 1940s. Then, when the TV companies realized that color TV was not only possible, but also way more interesting to watch, they started designing color TV equipment. The color TV we ended up with works just like black and white TV, but it has three electron guns, three camera tubes, and three times as much information to send. Here's why:

WHAT A BLAST!

Color Filters

Collect different colors of those plastic sheets you can wrap your school reports in. You can find them in stationery or hobby stores. Get a red, a green, and a blue.

Look through the red plastic. What you see is mostly. . . red (for cryin' out loud). All the other colors of light coming at your eyes are absorbed by the red tinted plastic. The green plastic makes everything look green. And the blue—it's just about all blue.

Now put all three filters together and look through them. How much can you see? Very little, I'll bet. Red, green, and blue filters absorb almost all visible light. See, only red light makes it through the red filter, but then that red light is absorbed by the green and blue. And the same goes for the other colors of light.

Now let's turn that around. Add all the light that gets through the red plastic *plus* all the light that gets through the green *plus* all the light that gets through the blue. The combination of red, green, and blue light gives us all the colors we started with. Scientists realized that all the tints that our eyes see can be created using these three colors.

So here's the deal: The light coming into a color TV camera is split up into three parts to go into three tubes. Each tube has a filter just like the tinted pieces of plastic. Therefore, one camera tube sees only red light, another sees only green, and the third tube only sees blue. The three colors of light are all sent separately to your TV at home, and each is shot out of a separate electron gun. The phosphors on the screen are set up in tiny groups of three: red, green, and blue again. So your whole color TV screen is made up of little dots in three colors. When you sit far enough away, those colors melt together into all the tints we know from real life.

TV Waves

Think about all this: Television signals go out just like radio signals. But television waves have to carry a lot more information than radio waves. They have to carry the picture as well as the sound. The picture is very complicated: 300,000 dots per picture; 30 pictures per second; 3 colors. We also have to have information in the signal that tells the television which line is which. Talk about a lot of stuff.

To get all that information out, a TV wave is very broad. The government has reserved a lot of high electromagnetic frequencies for TV. The picture is in an AM part of the signal, just like the information on an AM radio. And the sound is sent FM just like the sound on FM radio. And the information coming over the electromagnetic waves is very carefully set up to get the most into them as possible. In the U.S. we have a lot of channels that we can send through the air—83 right now. With all this, TV signals travel only about 150 miles.

Actually, TV signals travel a lot farther straight up. TV signals are at such a high frequency that they don't bounce off the stratosphere. They go right into space, like FM radio we talked about earlier. Think about that. (*Star Trek*—the original and the "Next Generation"—are zipping out into space at the speed of light, headed who knows where.)

Have you ever wondered how come TV channels start with 2? We used to have Channel 1 on the dial, but it turned out to be too close in frequency to some radio channels. You would hear the radio on TV! We were just trying to put too much information into the air.

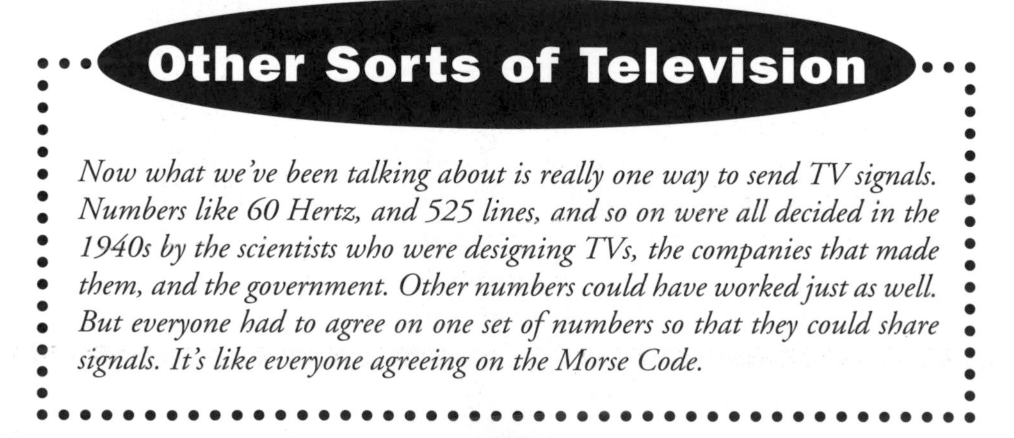

Other Sorts of Television

Now what we've been talking about is really one way to send TV signals. Numbers like 60 Hertz, and 525 lines, and so on were all decided in the 1940s by the scientists who were designing TVs, the companies that made them, and the government. Other numbers could have worked just as well. But everyone had to agree on one set of numbers so that they could share signals. It's like everyone agreeing on the Morse Code.

TV Overseas

TV sets in Europe are different from TVs here in North America. For one thing, the electricity blinks at 50 Hertz instead of 60 in many other parts of the world. In those countries where they have 50 Hertz electricity, they use 25 frames of TV a second. For another, instead of 525 lines on a TV screen, they use 625. Those systems were developed later, and they look a little sharper. That's why many foreign videocassettes don't work in our VCRs.

Cable and VCRs

We don't have to send TV signals through the air. We can also send those electronic signals over wires. That's how the programs travel from camera to director's booth to editing room, and so on. And if you have cable television or a videocassette recorder, then those signals are traveling through wires around your house.

We can also store those electronic signals on a videocassette. Inside every videocassette is a thin plastic tape covered with magnetic particles. The VCR records the TV signals in patterns on those particles.

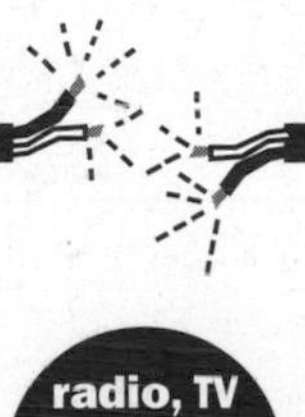

experiment

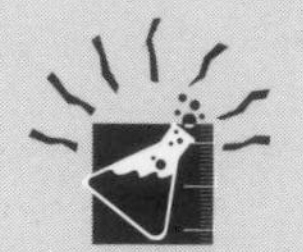

Technical Difficulties

Here's What You Need:

- *TV*
- *VCR*
- *Videocassette*
- *Magnet*

Here's What You Do:

1. *Tape something on the videocassette, something you don't need to save.*
2. *Take the cassette out of the VCR, and open its shield. Do this by pushing the little button on one side of the cassette and swinging the shield open.*
3. *Hold the magnet against the tape for a couple of seconds. This puts the particles on the tape inside the magnet's magnetic field.*
4. *Put the cassette back in the VCR and play it back with the sound all the way down.*

Here's What You See:

Where the magnet touched the tape, there's no program. That's because you placed the tape inside a magnetic field, and the particles lined up along that field's lines. Instead of being in a pattern that the VCR can read as a picture, the particles give you nothing, or static.

Satellite Dishes

You may have seen satellite dishes in people's yards or on the roofs of buildings. These are antennas that get television signals from satellites high above the earth. The dishes are so big because the amount of energy that gets to the earth's surface from a small satellite 36,000 km (23,000 miles) away is tiny. But with the right equipment, we can get the TV signal out of the air and watch it. Whoa.

The Next Generation

Scientists all over the world are working on better televisions, with brighter and crisper pictures. The latest thing right now is the High Definition TV that has 1000 lines of picture! With more lines, the picture looks sharper. Also, the way the information is carried in the TV signal is different and a little better. But to use these new TVs, everyone will have to agree on a new system, a new code.

Microwave Ovens

So how come I'm talking about microwave ovens in a chapter about radio and TV? Microwave ovens don't send information through space. They sure aren't very exciting to watch most of the time.

Microwaves are also part of the spectrum of **electromagnetic radiation**. Since "micro" means "very small," you can figure out that microwaves have very small wavelengths (about 1 to 100 cm). That means they have a very high frequency.

Microwave Popcorn

Here's What You Need:

- *Big spoonful of unpopped popcorn*
- *Glass that can go in the microwave oven*
- *Large glass bowl that can go in the microwave oven*
- *Water*
- *Microwave oven*
- *Hot pad*

Here's What You Do:

1. *Put the unpopped popcorn in the glass, and place the glass in the middle of the bowl.*
2. *Fill the bowl (not the glass) with water until the water is up to the level of the popcorn in the glass.*
3. *Put the bowl in the microwave oven and turn it on for two minutes. Does the popcorn pop?*
4. *Take the bowl out carefully with the hot pad. Is the water hot?*
5. *Pour the water out of the bowl. Put the bowl and the glass of popcorn back in the microwave oven just like before. Turn the oven on for two minutes. Does the popcorn pop this time?*

Here's What You See:

When the popcorn kernels are surrounded by water, they don't pop. Instead, the water warms up a little. Take out the water, and the kernels start to pop pretty quickly.

Microwave ovens work by heating up the water molecules inside them. In a way, water molecules are the "antenna" that catches the microwaves from microwave ovens. This radiation gets water molecules to vibrate. They start vibrating. And what do vibrating molecules mean? **Heat!** That's why the water in the bowl gets warm.

When there's no big puddle of water around to absorb the microwaves, they travel into the kernels of corn. Inside each kernel there's a little water. Those water molecules start to vibrate and turn into water vapor, which, you remember from Chapter 4, takes up a lot more space than liquid water—so it bursts open the kernel. That turns the kernel into a light, fluffy piece of popcorn.

Some microwave ovens tell you how much water they have to have inside in order to run without being damaged. If there isn't enough water to absorb the microwave energy, it is reflected back into the part of the oven that makes microwaves. That can hurt the machine. It's a good idea to keep a little container of water in a microwave oven even when it's turned off.

The Whole Spectrum

Here's a whole spectrum of electromagnetic radiation. Everything you see here—radio, TV, microwaves, cellular phones, light, heat, and more—comes in the same form of energy. That energy travels at the speed of light, and we can think about it coming in waves or in particles called photons.

Look for visible light, which has wavelengths of about 400 to 700 billionths of a meter, or frequencies of about 800 to 400 trillion Hertz. Pretty tiny range, isn't it? That's all our eyes can see. *All* this other energy is zipping around us at 300,000 km/second (186,000 miles/second), but we can't see it.

Electromagnetic Spectrum Chart

Type of Energy	Frequency	Wavelength
Electrical current's field	100 Hertz to 1 million Hertz	longer than 500 meters
Radio wave	500,000 Hz to 108 million Hz	600 m to 2.8 m
Television wave	55 million Hz to 800 million Hz	5.5 m to 37 centimeters
Microwave	1 billion Hz to 1 trillion Hz	30 cm to 3 millimeters
Infrared	1 trillion Hz to 430 trillion Hz	2 mm to 720 nanometers
Visible light	430 trillion Hz to 750 trillion Hz	720 nm to 420 nm
Ultraviolet	750 trillion Hz to 10 quadrillion Hz	420 nm to 10 nm
X-rays	10 quadrillion Hz to 10 sextillion Hz	10 nm to 1/500 nm
Gamma rays	100 quadrillion Hz to 10 sextillion	1/10 nm to 1/100 nm

WEATHER

chapter eight

Weather affects every living thing here on Earth, including you and me. What we wear, where we live, what we eat, and what we do all day depends on the weather. For the most part, weather is the interaction of air and water very near the Earth's surface. Now when we say very near, we mean the first 13 km (8 miles) or so. That's a long way up, but it's only the bottom layer of our **atmosphere** [ATT-muss-fear], which means "ball of air." The whole atmosphere is 550 km (340 miles) deep, though it doesn't really have an edge. The air just gets thinner and thinner and fades away into space.

The bottom layer of the atmosphere is called the "**troposphere**" [TROHP-us-fear]. "Tropos" means "turn" in Greek. So the troposphere is where things are turning. That's where all our weather takes place. Studying the weather is called "**meteorology**" [MEE-tee-or-OL-o-jee], which is Greek for "study of things in the air."

There are lots of things that help to create weather, but the three most important questions are:

1. **How much air is there in the area?** *(No kidding.)*
2. **How warm is the air?** *(What's the temperature?)*
3. **How much water is in the air?** *(Is it humid or dry outside?)*

The answers to those three questions help us scientists forecast thunderstorms, blizzards, sunny days, hurricanes, tornadoes, heat waves, and so on.

Our Atmosphere Sloshes Around

How much air is there? Isn't air around all the time? Of course—otherwise, we couldn't breathe. But sometimes there's *more* air overhead than others. When there's more air, meteorologists call it a "**High**," short for a "high pressure system." Less air is called a "**Low**."

High pressure systems are places in the troposphere where more air is piled up on the Earth's surface. If you could see from space, you'd see that the atmosphere is slightly thicker where the High pressure is. Remember that the atmosphere is held on the Earth by gravity, just like you and the ocean. So, if the atmosphere is sloshing in such a way that it gets deeper or thicker, then we'll experience higher pressure under the weight of the thicker air—a "High."

You can probably see right away that a "Low" is a place where the atmosphere is thinner—just slightly thinner. So, with less air over our heads, we feel slightly less pressure.

Meteorologists like you measure these pressure variations in the atmosphere with an instrument called a "**barometer**" [bare-AHM-mih-ter]. You can build one and measure "barometric" [BARE-uh-METT-trick] pressure. Try this:

e x p e r i m e n t

Building a Barometer

Here's What You Need:

- *A clean jar, like a pickle or jelly jar*
- *A balloon*
- *Drinking straw*
- *Rubber cement*
- *Ruler*
- *Tape*
- *Scissors*
- *Rubber band*
- *Long piece of wood*

Here's What You Do:

1. *Cut the neck off the balloon.*

2. *Paint the jar's rim with rubber cement.*

3. *Stretch the balloon over the mouth of the jar, and hold it flat with the rubber band.*

4. *Snip the end of the straw at an angle so that it forms a sharp pointer.*

5. *Tape the straw to the balloon so it sticks out to the side of the jar. Make sure the taped end is near the middle of the balloon, not the edge.*

6. *Using the rubber band and rubber cement, seal the balloon so no air can get past it into or out of the jar.*

7. *Tape the jar to a long piece of wood.*

8. *Stand a ruler against the wood near, but not touching, the straw. Tape it firmly to the jar so it can't slip.*

9. *Put the jar somewhere in your house where the temperature will stay the same: not near a heater or window.*

10. *Look at where the straw is pointing on the ruler. Write down that measurement and the date.*

11. *Come back the next day and look at the straw again. Is it pointing at the same mark on the ruler? If not, write down the new measurement and date.*

12. *If a storm is about to come through, look at the straw again. Write down where it's pointing this time.*

Here's What You See:

You'll see the tip of the straw move as the atmospheric pressure changes. When a storm comes through, the pressure in the jar will be a little higher than the pressure outside. The balloon will swell out a little, and the pointer will point to a lower number on your ruler.

The Italian physicist Evangelista Torricelli invented a barometer that uses a long tube of mercury. As the air pressure changes, the mercury goes up and down. It's very accurate. Nowadays we have several ways to measure air pressure that are related to Torricelli's device. Normal air pressure can be:

- **760 millimeters of mercury, also called 760.0 torr (named after Torricelli, shortened a bit)**
- **29.92 inches of mercury**
- **1013 millibars (1013 mb)**

EVANGELISTA TORRICELLI

1608-1647. Invented mercury barometer.

If a barometer measures how much air there is above you, what happens if you move up through the air? Check the next experiment:

Moving On Up

Use either the barometer you made in the last experiment or a manufactured barometer. Go to a tall building with an elevator, and ride up. You can see the barometric pressure go down. You are riding up through part of the atmosphere, leaving the densest, heaviest part below you. You can see the same changes in a car as you ride up and down tall hills. Not bad.

All Temperature

The temperature of the air is a measure of the heat energy in the air. You already know about heat from Chapter 4. But heat in the air affects the weather more than whether we feel hot or cold. See, differences in air temperature combine with differences in air pressure to make air move. It's what we call "wind."

Moving Air

First of all, the Sun shines on the Earth all the time. But it only shines on one side at a time: the daytime side. The other side, where it's night, is cooling off. But hey, what's a "side" of the Earth? The Earth is round and spinning, right? So the Earth is always being warmed and cooled. This makes the atmosphere expand and contract a little. One thing leads to another, and it gets windy.

Creating Wind

WARNING!

This experiment involves flame and hot light bulbs, so you need an adult to watch while you try it.

Here's What You Need:

- *A clean aquarium. If you can't find that, try a cardboard box with a piece of glass or plastic on one side.*
- *Heat lamp bulb or a spot light bulb, like the ones many people use for lighting outside their houses. If you can't get that, a candle is okay.*
- *Small bowl of ice*
- *Wax paper*
- *Matches*

Here's What You Do:

1. *Set up the heat lamp shining down in one end of the aquarium.*
2. *Put the bowl of ice at the other end.*
3. *Now you need to make a very small amount of smoke. Roll up some wax paper into a small tube about the size of a pencil. Light one end with a match, let it burn for a few moments, and then blow it out. It will make quite a bit of clean smoke.*
4. *Hold the smoking end of the wax paper in the aquarium near the bowl of ice.*

Here's What You See:

The air in the warm end of the tank will expand and rise. Then the air in the cool end will cool, contract, and sink. The cool air will flow in to replace the rising warm air and the warm air will be drawn over to replace the sinking cool air. A circulating wind will develop in your aquarium. And, you can see the wind by watching the smoke. Pretty cool, huh? This is a convection current, just like the ones we talked about in Chapter 4.

experiment

Which Way the Wind is Blowing

Do you want to tell where the wind is coming from? You don't need a weather vane; you just need a clean, wet finger. Wet your finger in your mouth and hold it straight up the air. Which side of your finger feels cold first? The side that faces the direction the wind is coming from.

You see, the air rushing over your finger makes the water molecules on your finger **evaporate**. And every evaporated molecule takes a little energy away from your finger. So you start to lose some heat energy on that side of your finger, and it feels a little cool.

This experiment also shows you what meteorologists call "**wind chill**." They realized that even though the temperature might be 10°C, on a very windy day it could *feel like* -5°C. That's because wind makes water in our skin evaporate, which makes us feel colder.

Water and Air

How does water get into the air? Well, the experiment about which way the wind is blowing gives us a clue. Water evaporates off fingers into the air. Of course, most of the water in the air evaporates off the oceans and other large bodies of water, not fingers. Each and every evaporating water molecule floats separately in the air.

Here's another way that heat energy affects weather: the hotter air is, the more water vapor it can hold. So on many hot days we feel "sticky."

The warm air already holds a lot of water, leaving no room for the water molecules in our sweat to evaporate and cool us off. On cold winter days some people use humidifiers to put a little water back in the air.

The amount of water floating in the air compared to all the water that air can hold is called "**humidity**" [hyou-MID-i-TEE]. It's written as a percentage. A humidity of 50% means that the air around you holds half of all the water molecules it can hold. Humidity of 100% means the air is holding all the water it possibly can, and some of that water starts to become liquid again. Scientists like you call that "**condensation**" [KON-den-SAY-shun].

experiment

The Water Cycle

Carefully measure out 250 mL (1 cup) of water into a big, wide bowl. Cover the bowl with clear plastic wrap. Then set the bowl in the sunlight. After a few hours, come back and look at the plastic wrap. It's wet with droplets of water. Where did that water come from? It came from **evaporation** and **condensation**. The liquid water molecules were carried away by the air molecules one at a time. Then, when they hit the plastic wrap, some of those water molecules stuck together and became droplets again. Shake all the droplets off the plastic wrap back into the bowl. Carefully remove the plastic so as to keep as much water as possible in the bowl. Then pour all the water into your measuring cup again. Do you still have 250 mL (1 cup) of water? It usually comes out very close.

The same thing is happening, on a much bigger scale, in the weather. Water from the ocean, from ponds and rivers, and from your sprinkler system and even from you, is evaporating into the air. In the upper troposphere, where the air is very cold, the water molecules turn back into liquid. For water vapor to turn back into a liquid, it needs a place to stick. (In the bowl and plastic wrap experiment, the water vapor stuck to the plastic.) High in the sky, water vapor sticks to dust and forms little droplets.

We know the bunches of water droplets in the sky better as "clouds." And when so many water molecules are stuck together that the air can't hold them up anymore, the water falls out of the sky. It's called "**precipitation**" [pree-SIP-i-TAY-shun] which means "stuff that falls."

From Out of the Sky

Precipitation usually happens along the line where a High and a Low are pushing on each other. Those lines are called "**fronts**." The changes in air pressure and temperature that happen along fronts cause the water in the air there to condense, and then a lot of different things can fall out of the sky:

Rain *Simple. Liquid water falling from the sky. "Acid rain" is the name for raindrops which contain chemicals from smokestacks and other sources of pollution. Acid rain is slightly more acidic than normal rain, so it changes the acid balance of lakes it falls into. That can cause some plants and animals to die out.*

Snow *Very small amounts of water (smaller than raindrops) freeze and stick together. Snowflakes have six sides because of the shape of water molecule.*

Freezing rain *This is rain falling onto ground that's so cold the rain immediately turns to ice. It makes for very slippery surfaces.*

Sleet *Raindrops can get so cold that they freeze on the way down. Because each bit of sleet is as heavy as a raindrop but made of solid ice, it stings when it hits you.*

Hail *Bits of sleet start to fall. The ice pellets are carried back up by strong rising air currents. They get covered with a new layer of liquid water, freeze again, and fall again. They're carried up again, and so on. If there's a hailstorm in your area, grab one of the bigger balls of ice. Break it open in your kitchen. You can even see the different layers of ice built up.*

Thunder and Lightning

Actually, we scientists should call this "lightning and thunder," because lightning always comes first. Lightning is a huge bolt of static electricity that has built up in the clouds coming down to Earth. This electricity is so strong that it can travel through the air, which normally is not a very good conductor. As the lightning zips through the air, it heats the air molecules, and they spread apart suddenly. That makes the huge "BOOM" we know as thunder.

You may know the trick that helps you figure out how far away the lightning storm is. Count the seconds between the lightning flash and the thunder. Sounds travels more slowly than light, so even though they start in the same place, the lightning reaches your eyes before the thunder reaches your ears. On a summer evening when the temperature is around 24°C (75°F), every three seconds between the lightning and thunder means the lightning was about 1 km (5/8 mile) away.

Dusty

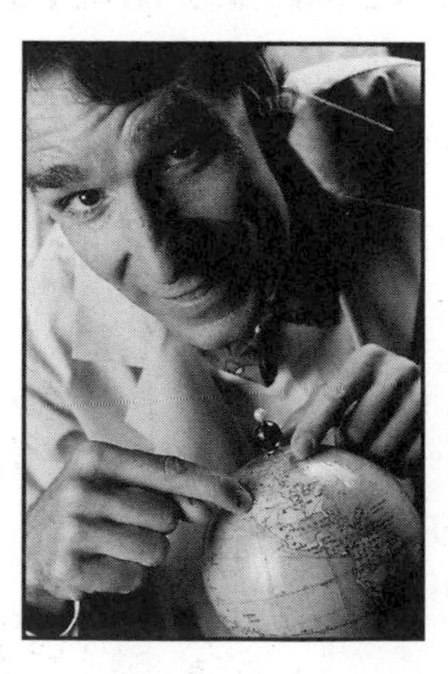

Every so often something besides air pressure, temperature, and humidity changes the weather. One of these things is a volcanic eruption. When a volcano erupts, it can throw a lot of dust into the air. That dust can block some of the Sun's heat, meaning temperatures go down. And water molecules stick to particles of dust to make rain more easily. Scientists believe that the eruption of Mount Pinatubo in the Philippines in 1990 caused the next couple of years to be a little colder and a little wetter in most of the Northern Hemisphere.

Predicting the Weather

It would be great to know exactly what the weather is going to be next weekend. Then we could be sure about scheduling picnics, or camping, or whether to get our skis out now or wait. Predicting the weather far ahead of time would mean that we could schedule school fairs only on days when it was sunny and warm. And we'd always know when to take a ski vacation!

Don't get your hopes up. It's not going to happen. Scientists have begun to realize that our atmosphere is swirling around so much that we're not going to be able to say exactly how it will come out on a particular day. Some meteorologists even say that if a butterfly flaps its wing somewhere in Africa, that can set up a whole chain of events that affects the weather on the other side of the world. Weather is a "**complex system**," meaning a very small change at the right place and time can have a very big effect later.

Meteorologists have to predict things in **probabilities**, like "50% chance of rain." That means, more or less, "On half the days which had weather just like this there was rain the next day, and on half of those days there was no rain." The farther into the future we weather scientists look, the more complex this predicting has to be, and the less certain we can be about any particular type of weather arriving on any particular day. So we'll probably never be able to predict the weather more than about five days in advance. When it comes to weather, there's always something to look forward to.

Where Does the Weather Come From?

Weather wouldn't happen here on Earth without energy from the Sun. That's what drives the winds, heats the air, and evaporates water from the seas. The Sun makes life on Earth possible because it keeps us warm. But the Sun just by itself wouldn't keep us warm. Take the Moon, for example. It gets hit with just about the same amount of energy from the Sun as we do, but it's not much of a place to live. When you're standing on the sunny part of the Moon, it's 130° C (266° F). If you're in the shade, it's about -180° C (-266° F). Why? Because there's no air to spread the heat out.

The Moon has a lot less gravity than we have here on Earth, only about a sixth as much. If you weigh 40 kg (100 pounds) here on your home planet, you'd weigh only

about 7 kg (17 pounds) on the Moon. That's all. Someday, take a look at the films of U.S. astronauts on the Moon. They're wearing huge massive space suits, but they're still jumping around as though they're on a trampoline. So what does this have to do with the weather? Well, the Moon's gravity is so low that it can't hold an atmosphere near its surface. And, without an atmosphere, there is no blanket of air to keep the surface warm.

The Greenhouse Effect

Heat from the Sun is absorbed by the Earth and everything on it: trees, rocks, raccoons, flowers, ferry boats, and you and me. After we get warm, we give some of the heat back off into the air. This process of objects giving off heat back into the atmosphere is what leads to what we scientists call the "Greenhouse Effect."

The idea behind a greenhouse, where gardeners grow lush plants all year long, is to hold heat in with glass. Here's what happens. All this heat radiation from the Sun comes into the greenhouse through the glass. Some of the energy gets absorbed by the plants, and some of it gets radiated again and bounces around inside the glass house. Most of that energy is reflected back into the greenhouse by the glass. See, ordinary glass lets visible light go through pretty easily, but it reflects most of the infrared light. So a greenhouse can keep itself warm in the day time.

Here's what happens on Earth: Some of the gases in the Earth's atmosphere act like greenhouse glass. The gases let energy from the Sun come in, but they reflect the heat energy that the Earth re-radiates right back toward us and the ground. These gases hold heat in! The most effective "**greenhouse gas**" seems to be carbon dioxide (chemical formula: CO_2). It's what we breathe out, and what our cars and trucks produce when they burn gasoline.

The Greenhouse Effect has been working here on Earth to keep us warm and cozy since primordial times. That's way back, perhaps 3.7 billion years ago. The problem now is that the Greenhouse Effect seems to be getting out of control. The human race is dumping about 20 billion extra tons of CO_2 into the air every year. Careful studies seem to show that the Earth is warming slightly with each year. If this trend continues, the Greenhouse Effect can go out of control. The polar ice caps may melt, the oceans will rise, and it will be too hot to farm the way we do now. Obviously, this could be very serious for all of us humans, not to mention other species of life.

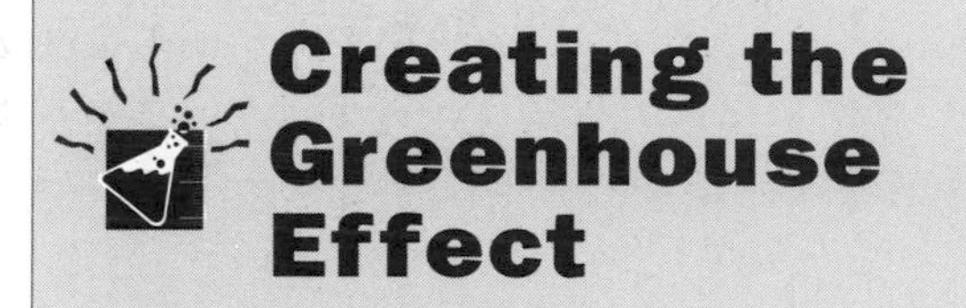

Creating the Greenhouse Effect

Here's What You Need:

- *An empty aquarium with a black or dark gravel floor*
- *Cardboard or plywood for a partition*
- *Tape or caulk*
- *Two identical thermometers that read the same at room temperature*
- *Baking soda*
- *Vinegar*
- *Large bowl*
- *Match*
- *Two identical bright lights, Make sure the bulbs use the same wattage.*

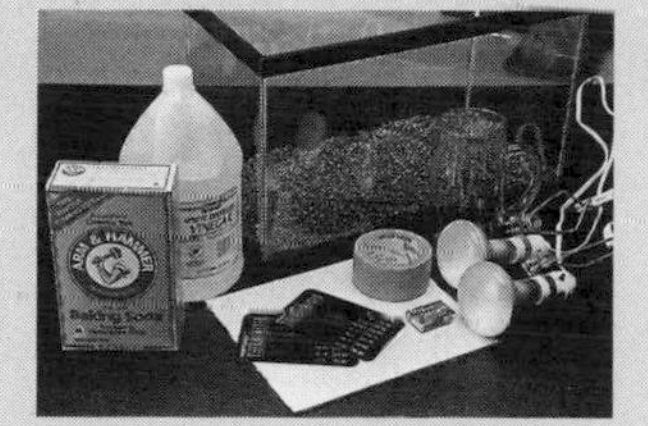

Here's What You Do:

1. *Build a partition in the middle of the aquarium out of the cardboard or plywood. Its edges have to fit pretty snugly to the sides of the aquarium.*
2. *Make the partition airtight by using a lot of tape or a strip of caulk. So now you have two separate but equal-sized glass chambers.*
3. *Place one thermometer in each chamber so that you can read them from outside. Shade them with a sheet of newspaper for a while.*
4. *Now fill one chamber with CO_2. Mix baking soda and vinegar in the bowl and place it in one chamber. You'll know the chamber is full when you strike a match that goes out as you slowly lower it to the chamber.*
5. *Shine a bright light at each chamber from equal distances. Wait about ten minutes, and then read the thermometers.*

Here's What You See:

CO_2 is heavier than air, so CO_2 will just sit in there for a long time, if it's not disturbed. After about 10 minutes the thermometer on the CO_2 side will be a few degrees warmer than the one on the plain air side. There you go, the Greenhouse Effect in your room.

The Rain Forest Factor

What do rain forests have to do with the Greenhouse Effect? People in countries that have rain forests want to catch up to the rest of the world. So they're cutting down their jungles and rain forests to make room for farmland. This increases the amount of CO_2 in the air in two ways. First of all, after they cut the trees down, they burn them. This pumps CO_2 into the sky. Second, the plants that were cut down consumed CO_2 and gave off wonderful life-giving oxygen. But they're gone—slashed and burned. So the destruction of the rain forests is only making the Greenhouse Effect worse.

You might wonder why people continue the destruction if it's so bad. People in these developing countries with rain forests see our life-style in North America, Europe, and Japan, and they want to live in many ways as we do. They believe that farmland will make them more money. They aren't convinced that the Greenhouse Effect is really a real thing. So they figure they'll just keep on cutting rain forests until something bad happens. Well, something bad is happening. Perhaps you will be among the scientists who come up with ways to preserve rain forests and provide rain forest native people with ways to enjoy the life-style they're looking for. It's a pretty important thing that we humans have got to work on.

The atmosphere not only keeps us warm, it protects us from the some of the Sun's rays that can hurt us. Namely, ultraviolet rays. Remember ultraviolet light from Chapter 5? That kind of radiation is just too energetic for us. Pale skin can handle only so much before it starts to burn, and our eyes are also vulnerable. Luckily, we're protected from this high energy sunlight by **ozone** [OH-zone] high in the sky.

OZONE LAYER

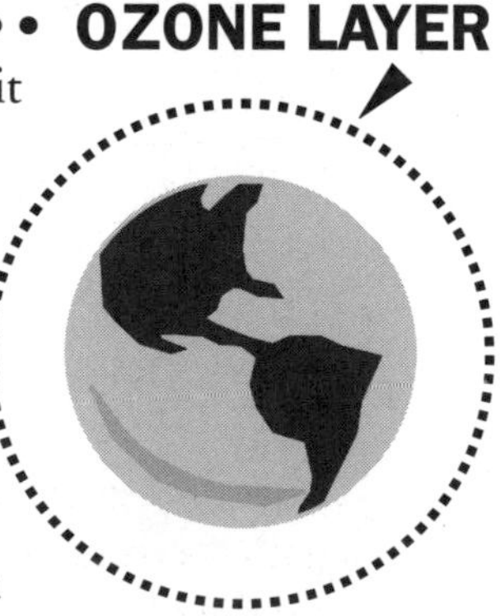

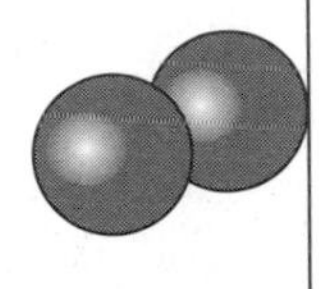

OXYGEN MOLECULE

Ozone is a special form of oxygen created by jolting regular air with energy, either electricity or ultraviolet light from the Sun. Oxygen usually comes in molecules of two oxygen atoms stick to each other. But ozone molecules have three oxygen atoms stuck together. The extra energy gives the atoms the push they need to bond in groups of three. An electric motor like the one in a toy train or slot car produces tiny, tiny amounts of ozone. You can smell the gas near these motors. (The word "ozone" comes from the Greek word for "smell.") The tiny amounts of ozone made by toys are not a problem. The large amounts made by cars, trucks, and buses are not good to breathe.

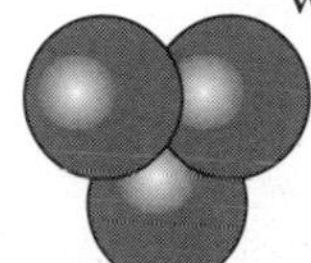

OZONE MOLECULE

Catch Some Rays

High up in the Earth's atmosphere, so high that the sky looks black instead of blue, regular oxygen molecules like the ones we breathe are hit with light from the Sun. Some of the oxygen atoms get knocked loose and reattach themselves to other oxygen atoms and form ozone. The first part of this process is called "**photolysis**" [FOE-tah-lih-siss], Greek for "knocked loose by light." So all day and night in the atmosphere, the Sun is jolting (photolyzing) some oxygen loose, which then re-bonds with other oxygen into ozone. Then, some of the ozone slowly breaks back down into regular oxygen.

Well, this process absorbs energy, which would otherwise shoot back down to Earth. The life here on Earth is not used to getting hit with this extra energy. If we do something to get rid of the ozone up high in the sky, all sorts of living things here on the surface—that's us again—could be messed up. And our crops, and the plants and animals that live here with us would be in trouble, too.

CFCs

About 20 years ago we discovered a problem with special chemicals called "**chlorofluorocarbons**" [KLOR-oh-FLOR-oh-KAR-bonz]. Sometimes people just call them "CFCs." They're chemicals made from the elements chlorine, fluorine, and carbon. We used them for years to clean things, make refrigerators and air conditioners work, and push paint out of spray cans. Anyway, CFCs make ozone break back down into regular oxygen very fast—too fast.

If there's less ozone in the upper atmosphere, too much of the Sun's energy may hit the Earth. That could upset the way crops grow, and the way animals and plants live. (Also, some people say that more people will get skin cancer if the ozone layer gets too thin. That's a minor problem. We can just keep our hats and shirts on.)

Here's the thing: CFCs break down ozone, but they don't break down themselves for a long time. CFCs act as a catalyst (a word from Chapter 2). They make the chemical reaction take place (ozone becomes oxygen), but the CFCs aren't affected much at all. So the same CFC molecules can hang out up in the atmosphere and mess up a lot of ozone before the Sun finally gets them to break down.

CFCs are breaking down the ozone layer and creating what some scientists describe as "**ozone holes**" over the Earth's North and South Poles. This could turn out to be very bad for all of us. We need to work on the problem right now. We need to stop letting all these CFCs into the air, and we need to find other chemicals that can do their job right away. Maybe you'll be the scientist who solves some part of these problems.

Why don't we make a lot more ozone and send it up? Cars, trucks, and busses burning gasoline all produce a chemical that turns into ozone when sunlight hits it. But ozone is a bad thing to have around down here on the Earth's surface. It's not good to breathe, and it wants to react with everything it comes in contact with and break it down. And we can't just pump our ground level ozone up to the upper atmosphere. It's just too far away and too spread out.

So we have two separate problems with ozone. First, we're destroying it high in the atmosphere. Second, we're creating too much of it here near the surface. Both problems are very bad for living things.

Being Sure

Are we absolutely sure about the Greenhouse Effect and the ozone holes?

Each of those ideas is a big hypothesis about how the chemicals in our atmosphere work, and remember that scientists always test and refine a hypothesis. There's still a lot we don't know about the weather. Scientists are just now starting to discover how plants react to and affect the atmosphere. For instance, a slightly higher air temperature in Alaska means the tundra [TUNN-dra] plants give off more CO_2, which in turn helps raise the temperature more. Yow!

We can't test a hypothesis by experimenting with the whole atmosphere. For one thing, it's too big. For another, it's the only thing we have to breathe! But scientists have done smaller experiments in labs, and done many observations of the atmosphere. They found a lot of signs that the Greenhouse Effect and the ozone holes really are big problems. If we ignore them now, we may not get a chance to fix them later.

KACHONG!

OUTER SPACE

chapter nine

Every night, since ancient times, men and women have sat and looked up at the stars. True, at night ancient guys and gals couldn't do a lot of the things we do. There were no lights, no books, no televisions, no microwave popcorn. But the night sky is a truly awesome sight.

If you live in the city, you can probably see only a few dozen stars because of all the light reflected from the ground. But if you look up at the sky away from the city's lights, you can see thousands of stars—thousands! It's breathtaking. It doesn't take you but a moment to feel very, very small.

See, in a way, we all live in outer space. Our planet is like a spaceship cruising through space. What makes "**astronomy**" [AH-strahn-OH-mee] so cool is that by studying the stars and outer space we learn so much about our own planet Earth and where we fit into the universe.

How the Universe Looks—at First

When our ancestors looked up at the sky at night, they assumed they were looking right at heaven. They thought the stars were on the "ceiling" of a great dome and that the Earth was flat. The Earth sure looks flat at first, especially when we look at a big smooth lake. Now, we all know it's round—we've seen pictures from space, for crying out loud. There's the Earth—round! Piece of cake, for us! But our ancestors didn't have that luxury. They had to figure it out for themselves.

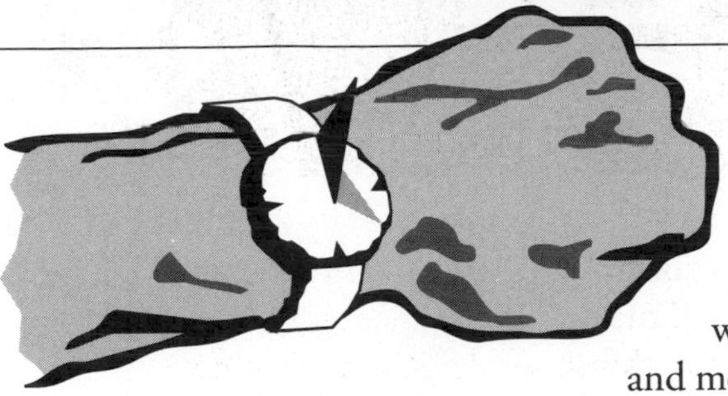

To start with, our ancestors watched the Sun. They kept track of the time of day so they knew when it would get too dark to walk around. Then they kept track of the days and months so that they would know when to plant their crops, when to harvest them, and when to have big wild parties. No kidding. Lots of ancient people had parties on the summer solstice [SOLE-stiss]—the longest day of the year. Some people around the world still hold big solstice parties.

The Round Earth

So most ancient guys and gals figured, "Well, the Earth is flat and the Sun goes over every day from east to west and somehow gets back to the east in the morning. Isn't that charming? Can't wait for the next solstice party." But a few ancient astronomers were busy figuring out the Earth was round. They were sure of it. They had two very important clues.

Every once in a while, we here on Earth get to see what's called a "**lunar eclipse**" [LOO-ner ee-klipps]. Lunar is Latin for "moon." Eclipse is Greek for "leave out." During a lunar eclipse the Earth gets between the Sun and the Moon and blocks the light that would normally hit the Moon. During some lunar eclipses, you can see the shadow of the Earth right on the Moon! Right up there. And you know what? The shadow is always curved.

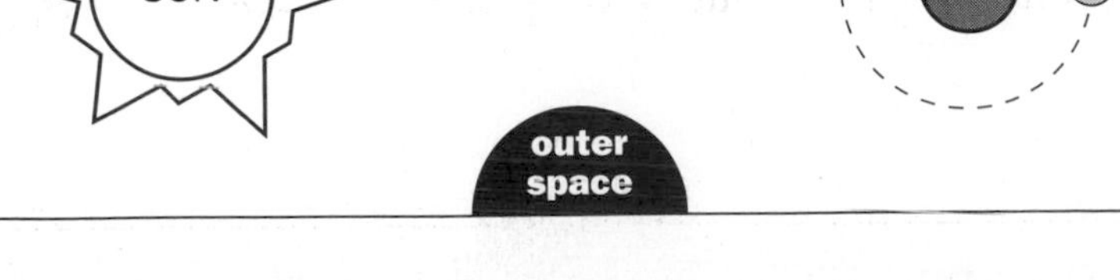

SUN

EARTH

MOON

Round Shadows

Shine a lamp on a wall. Then try to find an object that always casts a round shadow. How about a dinner plate? A football? A baseball cap? A basketball?

The ancient astronomers we're talking about here lived in Egypt. And, they were very good at geometry. They realized that the only shape that casts a curved shadow no matter which direction you shine light on it is a round ball—a **sphere**. Pretty cool thinking. They looked at the Moon when it was eclipsed, and realized they were living on a giant ball. Not bad.

Sinking Ships

Try this next time you're near the ocean. (If you live far away from water, don't forget this experiment; someday you'll have a chance to try it.) Watch a ship as it sails away from you or is moving along the horizon. If you can, use a pair of binoculars. Watch the ship as it goes over the horizon. If you look carefully, you'll see that part of the ship is below the line of the horizon! It seems like half of the ship is underwater, or has fallen off the edge of the Earth! How can this be?

Well, the Earth is round, and when the lower portion of a ship is past the farthest place you can see from where you're standing (the horizon), you only see the top part. This is particularly striking with tall sailing ships or large cruisers.

What must have people have thought? There's the ship—half gone! But, somehow, the ship doesn't fall off the edge into some weird Neverland. It just keeps sailing because the Earth, it turns out, is not flat—not at all.

Measuring the Earth

A few of those astronomers in Egypt were not only sure that the Earth is round, they even knew how big it is! This leads us to the extremely cool story of Eratosthenes [AIR-uh-TAHSS-thin-eez]. He was an astronomer living in Egypt almost 2200 years ago. He had read at the huge library in Alexandria very reliable reports that south of him in the town of Syene [SEE-nuh] the following thing happened: "At noon, on the longest day of the year, the Sun shines straight into our water wells; we can see all the way to the bottom!"

ERATOSTHENES

Around 350 B.C.
Measured the Earth using shadows.

Now Eratosthenes thought about this: Sun straight into wells. . . hmmm; no shadows at noon. . . hmmm; longest day of the year. . . hmmm. Then he tried it himself. On the longest day of the year, the summer solstice, he tried looking into wells in Alexandria. And he couldn't see the bottom.

So Eratosthenes thought, "Well, what's going on here? Why shouldn't people in Syene and Alexandria see the Sun the same way?" Eratosthenes realized then that the Earth is not flat—no way. It's round—it's a sphere. The Sun's light hits different parts of the Earth at different angles. At Syene the sunlight came almost straight down, so it could go all the way to the bottom of a well. At Alexandria, the Sun's rays came at an angle, so they hit a well's sides.

Being the way cool guy that he was, Eratosthenes also realized that he could measure the size of the Earth. First he measured the angle of the Sun's rays where he was, in Alexandria, with a shadow cast by a stick on the day of the solstice. Then he hired a guy to pace off the distance, as well as he could, to Syene. With this information he figured it all out.

When Eratosthenes measured the angle of the Sun's rays on the solstice in Alexandria, he found that it was about 1/50 of a circle. The walker told him it was about 830 kilometers to Syene. (Of course, they used ancient units of distance, not kilometers.) So our hero calculated that 830 km must be about 1/50 of the distance around the Earth. Therefore, the Earth must be: (830 km) x 50 = 41,500 km around. That's within about 4% of the modern number we use: 40,067 km (24,902 miles)! And we've got satellites and lasers and every doggone other thing to help us!

Do you see what Eratosthenes did? He measured the size of a whole planet! And he measured it very accurately! All because he was reading one day about sunshine and shadows, and he just got to thinking. That's some pretty big thinking.

Unfortunately, Eratosthenes's calculations were forgotten for centuries. By the time Christopher Columbus was sailing in 1492, nobody was sure how big the Earth was. Most sailors knew the Earth was round, not flat, but they thought it was too large to sail around. Apparently, even Columbus actually thought it was much smaller than it actually is. That's why he thought he was in India when he landed in the Caribbean, even though India is really 15,000 km (9,000 miles) away.

The Solar System

Now during all this time everyone assumed that the stars, the Sun, and the Moon were all going around the Earth—which, by the way, is wrong. They assumed it because that's the way it looks at first. By looking closely at the sky at night, they found five other things, too. These things looked a bit like stars, but instead of staying in the same place compared to other stars, they showed up in one place on one night, and a little bit away the next. Some even seemed to go one direction, stop, and go back the other way. The ancient astronomers called these objects "wanderers" or, in Greek, "**planets.**" They assumed the planets were going around the Earth, too.

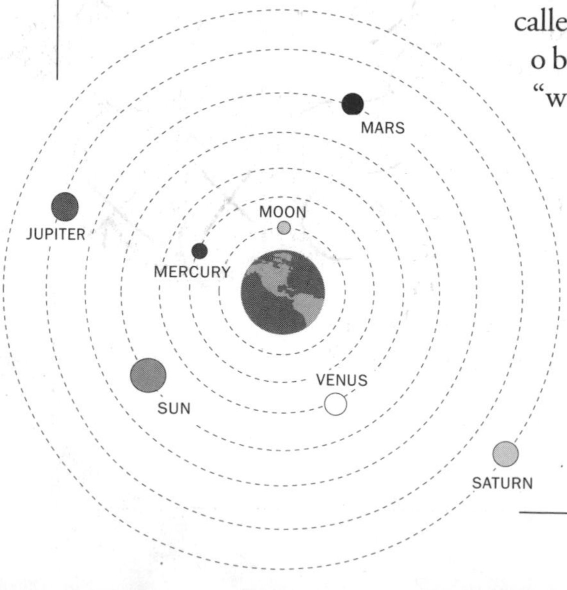

Another ancient astronomer from Alexandria named Ptolemy [TAHLL-im-ee] compiled many accurate observations of the stars and planets. Ptolemy's calculations were so good that people used them from around the year 200 all the way through Columbus's voyages. But after a few centuries Ptolemy's calculations started to have some problems. His predictions for where the planets would appear and when the Sun would come up started to be way off. See, he was predicting the positions of the planets using the idea that the Earth is at the center and everything goes around it. That hypothesis just wasn't working out.

Our Guiding Star

Well, this bothered a Polish mathematician named Nicolaus Copernicus. He reworked all of Ptolemy's calculations and realized that they came out better if he assumed that the planets revolve around the Sun, and if the Earth is a planet, too. So he reasoned that the Sun is at the center of the solar system (and probably the universe). This was a really wild idea at the time, about 1543. Copernicus was very nervous about publishing a book about his discovery because many people liked thinking about the Earth as the center of everything.

NICOLAUS COPERNICUS

1473-1543. Reasoned that Sun was center of solar system.

Copernicus changed the world. Over the next few decades scientists everywhere realized he was right. They now could really get down to thinking about our place in the universe. Historians like to call the publishing of Copernicus's book the beginning of "**The Scientific Revolution**." In other words, astronomy gave humans confidence that they could learn about the world. We could see where we fit in. Astronomy still has that cool feature about it today. Let's keep going.

So Far Away

In 1572, a very clever Danish astronomer named Tycho Brahe [TIE-koh BRAH-hee] watched a very bright star for many days and realized that its position never changed. He reasoned that the star must be very very far away. Much farther than people had thought.

TYCHO BRAHE

1546-1601. Observed planets and stars very accurately.

If you're riding a bicycle (or in Tycho's day, riding a horse), things which are close seem to whiz by. Zoom! There goes a tree! Zoom! There goes a house! But big things which are very far away, like a mountain on the horizon or the Sun, don't seem to move fast at all. Tycho looked up at this very bright star, and measured it with his best instruments, and discovered that it didn't seem to move at all, either. So it must be very very far away.

Elliptical Thought

The king of Denmark thought Tycho's discovery was great. "What cool reasoning by my man Tycho!" So his royal highness gave Tycho an observatory. Tycho watched the stars and wrote down their positions for years. Later his assistant, Johannes Kepler [YO-hahn KEPP-ler], studied the charts and positions extremely closely, just about painfully, and realized that planets don't go in circles around the Sun. Instead, the planets go in "ellipses" [ee-LIPP-siz]. You can draw an ellipse.

JOHANNES KEPLER

1571-1630. Discovered planets travel in ellipses around Sun.

experiment

Ellipse

Here's What You Need:

- *Thumbtacks*
- *60 cm (2 feet) of string*
- *Big piece of paper*
- *Pencil*
- *Bulletin board. If you don't have a bulletin board, find another flat surface that is okay to stick thumbtacks into. Do not, for example, use the dining room table.*

Here's What You Do:

1. *Tack the corners of one piece of paper to the bulletin board.*
2. *Tie the string in a loop. Stick a tack into the center of the paper, and hang the loop around it.*
3. *Put the pencil in the loop and pull the string tight. Keeping the string tight the whole time, move the pencil all around the tack. What shape do you draw? Okay, it's a circle.*
4. *Now, stick another tack into the paper about 3 cm (1 inch) away from the first one. Slip the loop around both tacks.*
5. *Put the pencil in the loop and pull the string tight. Keeping the string tight the whole time, move the pencil all around the tacks.*

Here's What You See:

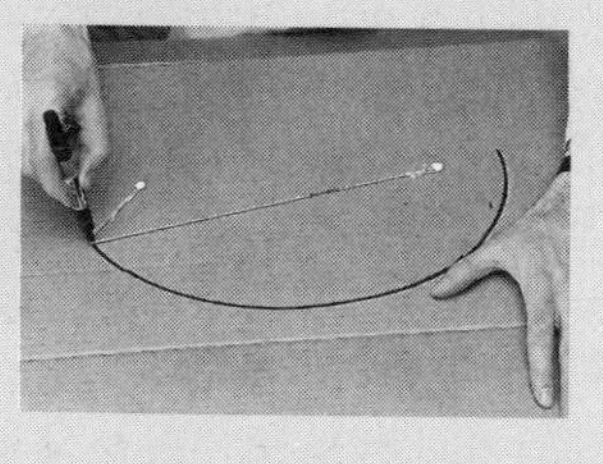

This shape drawn with a set loop of string around two points (called "**foci**" [FOE-sye]—the plural of "focus") is called an ellipse. Try drawing another ellipse, this time with the thumbtacks 10 cm (4 inches) apart. It's a lot flatter, right? See, an ellipse is like a circle only squeezed and stretched.

An ellipse is the type of path that all the planets follow around the Sun, and that the Moon follows around the Earth. The paths aren't quite circular after all. Kepler was the first human to see this, and he figured out that the Sun is at one of the thumbtacks in the last experiment. The Sun is at one **focus** of the ellipse. The other focus is just out in space; it's what a scientist (like you) might call a "theoretical" point in space. That's probably a better name than "**Other Thumbtack in Space**."

Kepler's ellipse turned out to be quite an insight because now there were no more errors in any of the predictions made for the paths of the visible planets. The planets were all exactly where Kepler and Tycho thought they should be. Quite a deal.

Satellite View

Another big discovery about the Solar System (meaning "belonging to the Sun") was made by Galileo. Remember how he may have tested gravity off the Leaning Tower of Pisa? Well, he was one of the greatest scientists of all time. He made a lot of discoveries. In astronomy he was the first scientist to point the new invention called the "telescope" up at the stars.

If you recall from Chapter 5, a telescope is two lenses put together to make distant objects look bigger. Galileo watched the planet Jupiter through his telescope, and discovered that there were four little dots of light spinning around it. He figured out that these were moons, just like the Moon around the Earth. Galileo also looked at Saturn, and saw what we later figured out were rings around it. He showed other scientists that there's a lot more out there in the universe than people had thought.

GALILEO GALILEI

1564-1642. Used the telescope to study the stars.

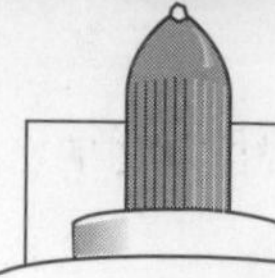

Holding Everything Together

The next mystery for astronomers to solve was what was holding this Solar System together. I mean, how did a planet keep whirling around the Sun with nothing between them? This was about the same time that Galileo was figuring out how gravity worked by letting things fall to the Earth. The English scientist Isaac Newton put these observations together and found the answer.

Newton was looking at the Moon and realized that it must be falling. The Moon is falling toward the Earth just like apples fall off trees, only much more slowly. Hmmm. He realized that gravity was pulling the Moon towards the Earth at the same time that its momentum from moving through space pushed it away from the Earth. And, because the Moon has been there since before people were around, the gravity and the momentum must be in balance. Here's how to see forces in balance:

ISAAC NEWTON

1686-1736.
Saw momentum and gravity hold planets in orbit.

experiment

Balanced Forces

Here's What You Need:

- *2 rubber balls of different weights*
- *8 m (24 feet) of string*
- *Paper towel tube*
- *Scissors*
- *Tape*

Here's What You Do:

1. *Tie one ball very carefully to the string. Secure the string with tape.*
2. *Run the string through the paper towel tube. Tie the second ball to the loose end of the string. Secure the end of the string with tape.*
3. *Take the balls and tube contraption and the scissors outside to a wide flat space where you can't break anything: a big backyard, a field, or an empty parking lot.*
4. *Hold the tube and start the lighter ball spinning over your head. Spin it so that the heavier ball hangs at a distance below the tube.*

5. *While the top ball is spinning, cut the string with the scissors. How do the balls react?*

Here's What You See:

When you cut the string, the spinning ball flies off in a straight line. The dangling ball falls straight down to the ground. The force of gravity pulling on the dangling ball was balancing the momentum of the spinning ball—until you cut the string.

In the Solar System, there's no string, but the Sun and planets are so massive that there's a lot of gravity. The Sun and planets pull on each other with gravity. That balances out the momentum of the planets zipping through space. What if the pull of gravity suddenly went away? Planets and moons would sail off away from the Sun. They probably won't: remember that gravity is one of the Four Fundamental Forces of the universe, and it seems to be permanent.

GRAVITY MOMENTUM

New Discoveries

The same balance of gravity and momentum holds the Moon in balance with the Earth. In fact, gravity holds every planet and every moon in the solar system in its orbit. Newton understood gravity inside and out. No joke. The laws of gravity that he discovered are so accurate that they were used to find a new planet. (A whole planet—that's a big thing.)

The first "new planet" was Uranus [nowadays we say YER-un-uss], discovered in 1781 by an astronomer named William Herschel. Now, Uranus had been there all along, but it was too dim for people to see. For thousands of years, people thought there were only six planets, including Earth. Using a telescope, Herschel found number seven.

Then astronomers started to watch Uranus and soon realized that its orbit was kind of weird. It wasn't following a smooth ellipse around the Sun. Using the laws of gravity, people figured out that there must be another planet out there near Uranus whose gravity is tugging

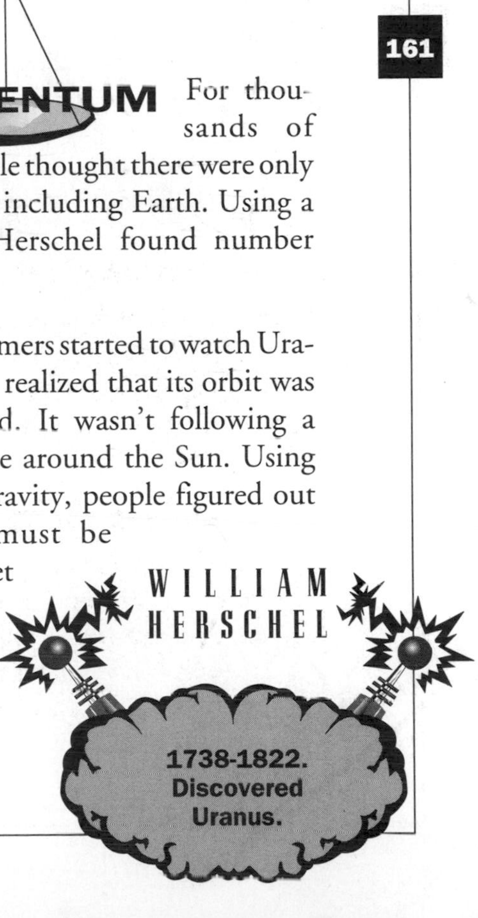

WILLIAM HERSCHEL

1738-1822. Discovered Uranus.

on its orbit. Sure enough, in 1846, sixty-five years after Uranus was found, astronomers found the planet Neptune [NEP-toon], right where Newton's laws said it should be. In other words, the laws of gravity are so accurate and reliable that they can be used to find objects in the sky 4 1/2 billion kilometers (2.8 billion miles) away from Earth. It's astonishing.

The orbits of the planets, their moons, the stars, and other celestial bodies, as near as we could tell, seem to run just like a clock. Everything shows up where it should right on time. This greatly influenced people's thinking, scientists and non-scientists alike, because it was clear that the universe and the world are pretty orderly. If we are careful, we can figure it out.

The Solar System

planet	moons (• major)	planet mass compared to the Earth's mass	distance from the Sun in kilometers
Mercury		6/100	58,000,000
Venus		82/100	108,000,000
Earth	• Moon	1	150,000,000
Mars	• Phobos • Deimos	11/100	228,000,000
Asteroid Belt—Asteroids are rocky objects orbiting the Sun that are much smaller than planets. There are about 100,000 asteroids, but all together they would weigh less than the Moon. Three very large asteroids are named Ceres, Pallas, and Vesta.			
Jupiter	Metis, Adrastea, Amalthea, Thebe • Io • Ganymede • Europa • Callisto *These four moons, each about as big as our Moon, are what Galileo saw with his telescope.* Leda, Himalia, Lysithia, Elara, Ananke, Carme, Pasiphae, Sinope	318	778,000,000

planet	moons (• major)	planet mass compared to the Earth's mass	distance from the Sun in kilometers
Saturn	*7 wide rings, which are only about 2 km thick!* Pan, Atlas, Prometheus, Pandora, Epimetheus, Janus, Mimas, Enceladus • Tethys Calypso, Telesto • Dione Helene • Rhea • Titan, *a moon which has an atmosphere* Hyperion, Iapetus, Phoebe	95	1,426,000,000
Uranus	*9 thin rings* Cordelia, Ophelia, Bianca, Cressida, Desdemona, Juliet, Portia, Rosalind, Belinda, Puck • Miranda • Ariel • Umbriel • Titania • Oberon	15	2,870,000,000
Neptune	*4 very faint dark thin rings* Naiad, Thalassa, Despina, Galatea, Larissa, Proteus • Triton, *which has a very thin atmosphere* • Nereid	17	4,497,000,000
Pluto	• Charon—*nearly the same size as Pluto*	2/1000	5,914,000,000
Oort Cloud of Comet Material			*100,000,000,000,000 (that's one hundred trillion kilometers away from the Sun!)*

In the Beginning. . . There was Dust!

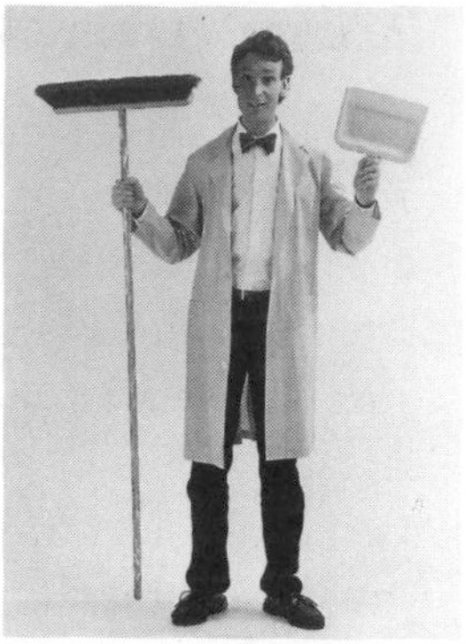

This brings up something else that is just astounding. All of the stuff that you and I and the Earth and the Moon are made of comes from stars. It comes out of the stars when they explode. Now that's something to think about!

Remember that stars are made of gas, including the electrically-charged gas called "**plasma**." Stars get their energy from the pressure of gravity (there it is again) pulling all that gas together. When the gas (hydrogen mostly) gets crushed together, it starts to explode. It explodes with the same type of energy we've created a few times here on Earth—namely, the energy released by a hydrogen bomb. The explosion in a star is continuous. It goes on all the time. The explosion is so huge that it sticks those hydrogen atoms together into atoms of other elements; this is called "**fusion**" [FEW-zhun]. The pressure of gravity is in balance with the pressure of the energy (heat and light) that the star gives off, so it keeps going.

Death of a Star

After a while—oh say, 15 billion years—the pressures become unbalanced, and the star collapses. Then the pressures build up again, and it explodes! The extremely bright star that our good buddy Tycho Brahe was looking at in 1572 was actually a star exploding—what scientists like you call a "**supernova**" [SOO-per-NOH-vuh], meaning "big new thing." When a star explodes, all the elements are formed. Right there in space. The stuff that we are made of goes zinging off in all directions and starts drifting through space.

The guy who worked this out, who came up with the theory of explosions that helps us understand the composition of stars, was Hans Bethe [BAY-tuh]. He got a Nobel Prize for it. With his calculations he was able to predict which elements should be found in the Earth's crust, and how much of each element we should find. He was right, to within a few percent. What a cool line of reasoning—star stuff.

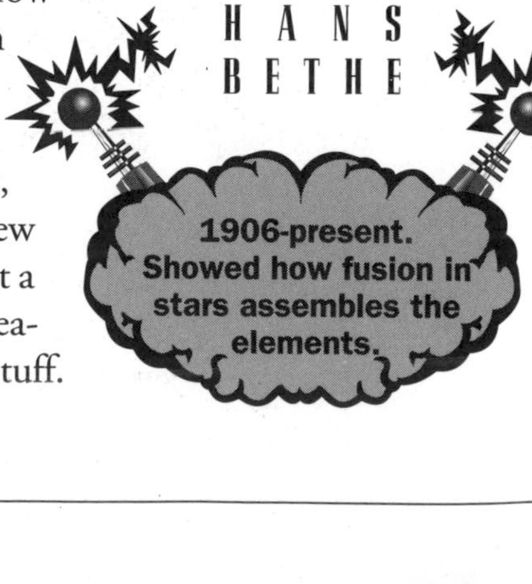

SUPERNOVA
BORN
15,000,000,000 BC
DIED
1572 AD

HANS BETHE

1906-present. Showed how fusion in stars assembles the elements.

Birth of a Star

Now, this stuff from the stars goes drifting through space. After a while, the dust's own gravity draws it together. If there's enough dust, it comes together to form a new star and new planets! First, because the dust comes in from all directions, it usually remains a cloud for a while. Then, as gravity packs the dust cloud in tighter and tighter, it ends up spinning. All the new planets, and the new star itself, are spinning. It's like an ice skater who goes faster and faster as she pulls her arms in—as she "compresses" her mass. Isn't that just, if I may say, out of this world?

The elements that we are made of—carbon, oxygen, sulfur, iron, and nitrogen—turn out to be the very elements that exploding stars make the most of. Can you even imagine? You and I and all living things are made of star dust! Whoa. The guy who saw how important the chemical elements are (and got me excited about it, and I hope you) is the very famous astronomer Carl Sagan. Whoa, whoa, whoa.

Collapsing Stars

When a star explodes, a lot of its material is sent shooting out into space. The rest is shoved into a smaller ball of matter where the star used to be. The gravity of this ball pulls the matter tighter and tighter, until the same amount of matter is in a much smaller space. This type of star is called a "**dwarf**." Obviously, dwarf stars are a lot smaller than the Sun and other regular stars. But they're just as heavy, sometimes even heavier. The matter is stuck together so tight it starts to behave in weird ways.

WHOA! WHOA! WHOA! WHOA!

Pulsars

In 1967 a British astronomer named Jocelyn Bell detected a new kind of signal from deep in space. It was a steady pulse of radiation, so she and her colleagues named this object a "pulsar." If it were visible light, it would look like "blink... blink... blink... blink..." day after day. Was some alien race trying to signal us?

JOCELYN BELL

1943-present. Discovered pulsars.

Pulsars

Here's What You Need:

- *Flashlight*
- *Strong string*
- *Bar or tree branch*

Here's What You Do:

1. *Tie the string around the flashlight. When you hold the string, the flashlight should hang horizontally. The string should hold the flashlight so securely that it can't slip.*

2. *Tie the other end of the string to the bar or branch.*

3. *Twist the flashlight around thirty or forty times.*

4. *Turn the flashlight on, let it go, and stand back.*

Here's What You See:

As the string untwists, it twirls the flashlight around in a pretty steady rhythm. When you stand back, you see the light from the flashlight in rhythmic flashes. Every time the flashlight bulb is pointed your way, you see light. The rest of the time, nothing.

Scientists realized that pulsars sent out signals in the same way that the twirling flashlight sends out light. The flashlight and pulsar are burning all the time, but only in one direction. The pulsar and the flashlight on the string are also rotating. Every time the star "blinked," it was pointed in our direction. But it was sending out the same "signal"—just a steady stream of energy—all the time.

Pulsars are a type of star that is under huge pressure. Much more pressure than we've ever felt on Earth. So much, in fact, that it overcomes the strong atomic force and shoves the electrons in the star's atoms into the nuclei. That kind of star is called a "neutron star," since it seems to be made entirely of neutrons.

Some collapsed stars have too much mass even to become neutron stars. No one has ever seen such stars, however. That's because, according to this hypothesis, they become "black holes," with so much gravity that they don't letlight or any other sort of energy escape. That's almost unbelievable. But astronomers have spotted stars out in space that are spinning around something almost unbelievably heavy and dark. That could be a black hole. It's like something a science-fiction writer might have made up, but it seems to be real!

Black Hole

Galaxies

For a long time, people have realized that the Sun is a star. It's just like all the other stars we see at night. Well, not like all those stars. Many of those glowing dots that look like stars are actually thousands of stars clumped together very far away. Scientists (like you) call that a "galaxy" [GAL-ak-see]. All the thousands of stars orbit around the center of the galaxy. Scientists aren't sure what that center is—one hypothesis is that it's a black hole.

The Sun is part of a galaxy, too. Since we're in the middle of that galaxy, we see it as a stripe of sky where the stars are thicker called the "**Milky Way**." If you go into the country on a clear night, you may be able to see a milky white glow—a band of star haze in the sky. It's the glow of the other stars in our galaxy! We are in a giant (I mean hugely, gigantically giant) galaxy of stars. It's a disk, and we are kind of out toward the edge. We see the disk edge-on; that's why it looks like a stripe across the sky. We are living right by a typical star, in a typical galaxy.

Speeding

Now here's a fact that is really wild: Everything in the universe is moving at a tremendous rate of speed. Right now you are on a planet that spins around every 24 hours. That means if you stand on the Equator, you're moving more than 1,730 km/hour (over 1,000 miles/hour)! The Earth is also moving around the Sun at about 108 km/ hour (66.8 miles/hour).

Furthermore, the Sun is moving, too, and it's carrying the Earth along with it. The Sun is spinning around the center of the Milky Way galaxy at about 250 km/ second (155 miles/second). And the Milky Way is also moving away from the other galaxies. No need to tie the furniture down, though: these motions are almost perfectly constant, and they have been going on for a long time. So they feel normal. We can't even notice them.

Do you see why astronomy is so important? It's what makes us aware of how the universe works. It tells us our place in the universe. Astronomy is the scientific subject which is at the forefront of human understanding. It is just the coolest. We need to keep exploring space, to keep looking out toward the stars.

Understanding the weather on our neighboring worlds, like Mars and Venus, may help us to understand more thoroughly the Greenhouse Effect, and save our planet. Observing the radio waves coming from other galaxies may tell us something about how we can make energy in the future for our society. Astronomy allows us to ask these questions and to study them. It's a super science that's been with us since the beginning of our existence on this planet, and will guide us as we continue to explore this amazing universe.

WHAT A BLAST!

INDEX

U

V

W

Z